G000167804

MOBILE MULTIMEDIA:
COMMUNICATION ENGINEERING PERSPECTIVE

MOBILE MULTIMEDIA:
COMMUNICATION ENGINEERING
PERSPECTIVE

ISMAIL K. IBRAHIM AND DAVID TANIAR
EDITORS

Nova Science Publishers, Inc.

New York

Copyright © 2006 by Nova Science Publishers, Inc.

All rights reserved. No part of this book may be reproduced, stored in a retrieval system or transmitted in any form or by any means: electronic, electrostatic, magnetic, tape, mechanical photocopying, recording or otherwise without the written permission of the Publisher.

For permission to use material from this book please contact us:
Telephone 631-231-7269; Fax 631-231-8175
Web Site: http://www.novapublishers.com

NOTICE TO THE READER

The Publisher has taken reasonable care in the preparation of this book, but makes no expressed or implied warranty of any kind and assumes no responsibility for any errors or omissions. No liability is assumed for incidental or consequential damages in connection with or arising out of information contained in this book. The Publisher shall not be liable for any special, consequential, or exemplary damages resulting, in whole or in part, from the readers' use of, or reliance upon, this material.

This publication is designed to provide accurate and authoritative information with regard to the subject matter covered herein. It is sold with the clear understanding that the Publisher is not engaged in rendering legal or any other professional services. If legal or any other expert assistance is required, the services of a competent person should be sought. FROM A DECLARATION OF PARTICIPANTS JOINTLY ADOPTED BY A COMMITTEE OF THE AMERICAN BAR ASSOCIATION AND A COMMITTEE OF PUBLISHERS.

LIBRARY OF CONGRESS CATALOGING-IN-PUBLICATION DATA
Available upon request

ISBN 1-60021-207-7

Published by Nova Science Publishers, Inc. ✦ New York

CONTENTS

PREFACE

Market Studies indicate that the number of subscribers for mobile communications has increased much faster than predicted, particularly for terrestrial use. In the year 2000, the number of mobile subscribers was approximately 400 million worldwide, and in the year 2010 more than 1.8 billion mobile subscribers are estimated. An interesting fact was presented in a new report by Telecommunications Management Group, Inc. (TMG) providing the statistical basis to show the number of mobile multimedia users exceeding 100 million in 2004. This breathtaking fact inspires us to start researching the mobile multimedia in all possible related aspects.

Research in mobile multimedia is typically focused on bridging the gap between the high resource demands of multimedia applications and the limited bandwidth and capabilities offered by state-of-the art networking technologies and mobile devices.

Communication engineering approaches this problem by considering not only characteristics of the networks and devices used, but also on the tasks and objectives the user is pursuing when applying/demanding mobile multimedia services and exploit this information to better adapt those services to the users' needs. This method is referred to it as user centric multimedia processing approach.

OVERVIEW OF MOBILE MULTIMEDIA COMMUNICATION

Mobile multimedia can be defined as a set of protocols and standards for multimedia information exchange over wireless networks. It enables information systems to process and transmit multimedia data to provide end users with services from various areas, such as mobile working place, mobile entertainment, mobile information retrieval and context-based services.

Multimedia information as combined information presented by more than one media type (i.e. text, pictures, graphics, sounds, animations, videos) enriches the quality of the information and is a way to represent reality as adequately as possible. Multimedia allows users to enhance their understanding of the provided information and increases the potential of person to person and person to system communication.

Mobility as one of the key drivers of mobile multimedia can be decomposed into:

1 *User mobility*: The user is forced to move from one location to location during fulfilling his activities. For the user, the access to information and computing resources is necessary regardless his actual position. (e.g. terminal services, VPNs to company-intern information systems).

2 *Device mobility*: User activities require a device to fulfill his needs regardless of the location in a mobile environment (e.g. PDAs, notebooks, tablet pc, cell-phones, etc).

3 *Service mobility*: The service itself is mobile and can be used in different systems and can be moved seamlessly among those systems (e.g. mobile agents).

The special requirements coming along with the mobility of users, devices, and services and specifically the requirements of multimedia as traffic type bring the need of new paradigms in software-engineering and system-development but also in non-technical issues such as the emergence of new business models and concerns about privacy, security or digital inclusion to name a few.

For instance, in the context of mobile multimedia, 3G communication protocols have great deals. Even some mobile protocol expert tends to define 3G as a mobile multimedia, personal services, the convergence of digitalization, mobility, and the internet, new technologies based on global standards, the entire of the aforementioned terms. In 3G, end user will be able to access the mobile internet at the bandwidth at various bit rates. This makes great challenging for handset device manufacturers and mobile network operators. In addition, large number of application and related issue need to be addressed considering the heterogeneity nature of the Internet. As well as, the various and rich content of the internet should be considered whenever one of the roles of 3G is being deployed.

In network traffic point of view, the majority of traffic is changing from speech-oriented communications to multimedia communications. It is also generally expected that due to the dominating role of mobile wireless access, the number of portable handsets will exceed the number of PCs connected to the Internet. Therefore, mobile terminals will be the major person-machine interface in the future instead of the PC. Due to the dominating role of IP based data traffic in the future the networks and systems have to be designed for economic packet data transfer. The expected new data services are highly bandwidth consuming. This results in higher data rate requirements for future systems.

BUSINESS DRIVERS

The key feature of mobile multimedia is reaching customers and partners, regardless of their locations and delivering multimedia content to the right place at the right time. Key drivers of this technology are on the one hand technical and on the other business drivers.

Evolutions in technology pushed the penetration of the mobile multimedia market and made services in this field feasible. The miniaturization of devices and the coverage of radio networks are the key technical drivers in the field of mobile multimedia.

1 *Miniaturization*: The first mobile phones had brick-like dimensions. Their limited battery capacity and transmission range restricted their usage in mobile environments.

Actual mobile devices with multiple features fit into cases with minimal dimensions and can be (and are) carried by the user in every situation.

2   *Vehicle manufacturer*: Furthermore, mobility also calls for new type of services (and thus revenues). Vehicle manufacturers want to improve the ears' Man- Machine Interface by using superior input/output devices. An open application platform would allow upgrading of multimedia equipment during the lifecycle of a vehicle, which is much longer than the lifetime of computing equipment. Safety equipment for automated emergency- and breakdown calls brings in positioning hardware into the car and thus enabling other location aware services. But vehicle makers also seek an after-market relationship to their customers: Once having ears connected, they can offer ear-specific services, including direction-finding and safeguarding support.

3   *Radio Networks*: Today's technology allows radio networks of every size for every application scenario. Nowadays public wireless wide area networks cover the bulk of areas especially in congested areas. They enable (most of the time) adequate quality of service. They allow location-independent service provision and virtual private network access.

4   *Mobile terminal manufacturers*: Mobile terminal manufacturers serve   individual people instead of households. Since there are more individuals than households the market is naturally bigger then the home terminal market. Furthermore, there is a large potential for use and fashion based diversification and innovation of terminals.

5   *Market evolution*: The market for mobile devices changed in the last years. Ten years ago the devices have not been really mobile (short-time battery operation, heavy and large devices) but therefore they have been expensive and affordable just for high-class business people. Shrinking devices and falling operation- (network-) costs made mobile devices to a mass-consumer-good available and affordable for everyone. The result is a dramatically subscriber growth and therefore a new increasing market for mobile multimedia services.

6   *Subscribers*: Persons, spend good percentage of their lifetime traveling, either for business or leisure, while they want to stay connected in every respect. This desire is more than proven by the current sales figures for mobile phones and the emerging standards for mobile narrow-band data services.

7   *Service Evolution*: The permanent increasing market brought more and more sophisticated services, starting in the field of telecommunication from poor quality speech-communication to real-time video conferencing. Meanwhile mobile multimedia services provide rich media content and intelligent context based services.

8   *Vehicle terminal manufacturers*: Vehicle terminal manufacturers currently suffer from vertical markets due to high customization efforts for OEM products. An open application platform would help them to reduce development time and costs. It is also a key driver for after market products. A wide range of services will increase the number of terminals sold.

9   *Ears*: For ear drivers, security and travel assistance are important aspects as well. They probably want to use the same services in the car they are used to at home and in the office. This is only possible with an open application platform.

Technology drives mobile multimedia with new means to communicate, cache, process, and display multimedia content:

- *Connectivity*: New means to communicate enable new services to be provided.

- *Memory and persistent storage*: developing memory technology allows caching of more content and offline processing, thus creating the illusion of instant access to interactive remote content. For example, audio/video content and whole websites may be downloaded in background and consumed offline. This is above all important for data broadcast services, which are transmitted in different fashions.

- *Processing*: More processing resources with less power consumption allow rendering of more complex multimedia content.

- *Display*: Visualizing multimedia content demands for cheap high resolution displays that comply to "handset" requirements "Screen driver, pixel per bits, width, height".

The value chain of mobile multimedia services describes the players involved in the business with mobile multimedia. Every service in the field of mobile multimedia requires that their output and service fees must be divided to them considering interdependencies in the complete service life-cycle.

1 *Network operators*: They provide end-users with the infrastructure to access services mobile via wireless networks (e.g. via GSM/GPRS/UMTS). The network operators want to boost the sales of network bandwidth, by enabling new types of services with new network technologies. In many countries a close cooperation between the transmitter and cellular network operators has been established to be able to offer hybrid network capacity. Service and content providers see the opportunity to promote and sell their services to people everywhere and anytime, thus increasing the total usage of their services.

2 *Content Provider*: Content provider and –aggregators license content and prepare it for end-users. They collect information and services to provide customers with convenient service collection adapted for mobile use. In another hand, some national broadcasters are forced by law to provide nationwide TV coverage through terrestrial transmission. They are interested in digital TV, because, firstly, it strongly reduces their transmission costs per channel by high ratio and, secondly, they improve the attractiveness of terrestrial reception which decreased strongly since the beginning of cable and satellite services

3 *Fixed Internet Company*: Those companies create the multimedia content. Usually they provide it already via the fixed Internet but are not specialized on mobile service provisioning. They handle the computing infrastructure and content creation.

4 *App Developers and device manufacturers*: Thy deliver hard- and software for mobile multimedia services and are not involved with any type of content creation and delivering.

ABOUT THIS BOOK

This book provides an insight into the field of Mobile Multimedia from a communication engineering perspective. The book is intended for people interested in mobile multimedia at all levels. The primary audience of this book includes students, developers, engineers, innovators, research strategists and IT-managers who are looking for the big picture of how to integrate and deliver mobile multimedia products and services.

While the book can be used as a textbook, system developers and technology innovators can also use it, which gives the book a competitive advantage over existing publications.

Despite the fact that mobile multimedia is the next generation information revolution and the cash cow that presents an opportunity and a challenge for most people and businesses. The book is intended to clarify the hype, which surrounds the concept of mobile multimedia through introducing the idea in a clear and understandable way. This book will have a strong focus on mobile solutions, addressing specific application areas. It gives an overview of the key future trends on mobile multimedia including UMTS focusing on mobile applications as well as on future technologies. It also serves as a forum for discussions on economic, political as well as strategic aspects of mobile communications and aims to bring together user groups with operators, manufacturers, service providers, content providers and developers from different sectors like business, health care, public administration and regional development agencies, as well as to developers, telecommunication- and infrastructure operators, ...etc.,

The book is organized into four parts. The introduction part, which consists of two chapters introduces the readers to the basic ideas behind mobility management and provides the business and technical drivers, which initiated the mobile multimedia revolution. Part two, which consists of six chapters, explains the enabling technologies for mobile multimedia with respect to data communication protocols and standards. Part three contains two chapters and is dedicated for how information can be retrieved over wireless networks whether it is voice, text, or multimedia information. Part four with its four chapters will clarify in a simple way how scarce resources can be managed and how system performance can be evaluated.

This book has been compiled from extensive work done by the contributing authors, who are researchers and industry professionals in this area and who, particularly, have expertise in the topic area addressed in their respective chapters. We hope the readers will benefit from the works presented in this book.

ACKNOWLEDGMENTS

Editors would like to acknowledge the help of all involved in the collation and review process of the book, without whose support the project could not have been satisfactorily completed. A special thanks goes to the Nova Science Publishers, Inc. Special thanks goes to Laurence T. Yang, the Embedded and High Performance Computing series editor, Frank Columbus and Maya Columbus, Department of Acquisitions, and Susan Boriotti, Senior Editor whose contributions throughout the whole process from initial idea to final publication, have been invaluable. We would like to express our sincere thanks to our employers Johannes Kepler University Linz, Austria and Monash University, Australia and our colleagues for supporting this project. In closing, We wish to thank all of the authors for their insights and excellent contributions to this book, in addition to all those who assisted in the review process.

Ismail Khalil Ibrahim

David Taniar

April 2006

In: Mobile Multimedia: Communication Engineering ...
Editors: I.K. Ibrahim and D. Taniar pp. 1-15

ISBN: 1-60021-207-7
© 2006 Nova Science Publishers, Inc.

Chapter 1

INVESTIGATION INTO THE MOBILITY MANAGEMENT MECHANISMS IN WIRELESS NETWORKS

Irfan Awan[*]
Department of Computing, University of Bradford,
Bradford BD7 1DP, United Kingdom
Muhammad Younas[**]
Department of Computing, Oxford Brookes University
Oxford OX33 1HX, United Kingdom

Abstract

Current developments in mobile computing and wireless networks have caused a significant transition from conventional enterprises to electronic enterprises. Enterprises provide customers with new opportunities and services for conducting business transactions through small handheld devices such as PDAs and mobile phones. These services are envisioned as the most convenient way of conducting businesses. Using such services from mobile devices equipped with GSM/GPRS, however, demands effective mobility management mechanisms in cellular networks - a popular architecture for wireless networks. Success of such services will largely depend on a reliable connection for the roaming users given the fact that mobile devices are characterized by low-bandwidth communication and limited processing power. Challenges posed to researchers are: to devise cost effective algorithms for modelling scarce channels in cellular networks, to maximize channel utilization, and to provide efficient handover mechanisms for roaming users. This chapter presents a framework for the performance evaluation of mobility management in cellular networks in order to maximize the channel utilization and to improve the handling of handover requests. The proposed framework is based on decomposition of cellular networks into individual cells and analyse each cell in turn as a queuing system with an enhanced priority mechanism. Typical numerical experiments are presented to demonstrate the performance of the proposed framework under various parameterisation settings.

---

[*] E-mail address: i.u.awan@bradford.ac.uk
[**] E-mail address: m.younas@brookes.ac.uk

1 Introduction

The realization of the potential of mobile communication has become evident in our daily life. It can be seen by the increasing use of mobile phones and other mobile devices in social, entertainment, and commercial activities. Further evidence of the popularity of mobile communication is the recently held 2006 3GSM World Congress (http://www.3gsmworldcongress.com/) which has attracted 50,000 visitors (including all major mobile technology vendors) from across the world. The aim of 3GSM was to further promote mobile communications and its applications into business using advanced mobile communication technologies. One of the fundamental technologies of mobile communication is the *wireless cellular networks* [1] which provides a roaming environment wherein users equipped with mobile devices can freely move from one cell to another. Such networks are constructed by dividing a geographic area into small service cells that support operations on distinct frequencies (channels) [2]. These networks have the advantage of maximizing the utilization of scarce frequency spectrum. Each mobile user communicates through a transmitter, called Base Station (BS), which then connects to the cellular network [3].

This chapter studies the mobility management mechanisms in wireless cellular networks in order to investigate the performance and reliability aspects of mobile services. We present a simple but an expressive example in order to identify the challenges of ensuring efficient and reliable mobile services. For example, a user may wish to travel to London to visit tourist attractions. While on the way to London she wants to buy tickets with her PDA in order to visit Madame Tussauds museum and to watch a theatre show in Royal Albert Hall. Such request may last for longer duration as it involves different services (e.g., museum booking service, theatre booking service, and payment service) which may be accessed via different communication lines and different computing devices. Thus she may issue a request from one cell and roam to another cell while her request is still in progress.

The above example demonstrates various requirements from the perspectives of users and the service providers. User's perspective is that services should be highly available, have good performance, and possess high reliability. Service provider's requirements are that their services should be accessed by a large population of users in order to maximize their profit and to gain the confidence of users in using their services. Service providers must offer reliable and efficient services to keep their businesses alive in the current competitive market. To achieve such reliability and efficiency, it is of paramount importance to provide users with continuous network connections during the processing of their requests. They must ensure that requests are handed over from one cell to another without losing data packets.

In wireless networks, roaming of user's requests is managed by network based handover control mechanism. When a user moves from one cell to another, this mechanism redirects user's requests at an appropriate moment to the destination mobile nodes. However, there are situations wherein no channels are available during the handover process. This results in the blockage of requests. A frequent handover of requests can cause network overhead. Also, if the requests are delayed for too long then they may be forcefully terminated. Blockage of the ongoing requests may have severe consequences for the service providers as it affects the performance and reliability of services and consequently their businesses.

The above discussion reveals that there is a greater need for efficient handover schemes that reduce request blocking. This chapter presents a framework for the performance evaluation of mobility management in cellular networks in order to reduce handover request blocking at a minimum network overhead. The proposed framework is based on enhanced priority queuing handover mechanisms with a buffer threshold — wherein a buffer consists of two partitions. The first part (before threshold) is shared by both handover and new requests and the second part (after threshold) is used to buffer only the handover requests. High priority is always given to handover requests through the Head of Line (HoL) queuing discipline. This framework provides user's requests with a seamless connectivity (as in above example). The system is modelled as a GE/GE/C/N/HoL queuing system with generalised exponential (GE) arrival and service time distributions, C channels, a finite capacity represented by a vector **N** and HoL priority scheduling rule. GE distribution is used to model the burstiness of the traffic.

The rest of the chapter is organized as follows. Section 2 reviews related work. Section 3 presents the design criteria for the proposed framework. Experiments and results are presented in Section 4. These results demonstrate that the proposed framework ensures low loss probability and mean Response Time for handover requests. Section 5 concludes the chapter and sets direction for future work.

2   Related Work

The process of request handover greatly affects the performance and reliability of wireless communication. It has therefore attracted significant attention from research [4-8].

The two classical handover mechanisms are: Channel Reservation and Queuing of handover requests [9]. Channel Reservation Handover reserves a set of channels for the handover requests. As in the above example, if a user has booked her museum ticket and if she moves to a new cell, then her request will be handed over to the reserved channel in the destination cell so as to complete the remaining part of her request (i.e., buying a theatre ticket). This technique introduces handover priority over new requests and decreases handover blocking probability. However, there are problems associated with this scheme. For instance, Zhuang et al [10] report that this scheme wastes the reserved channels in situations where the number of handover requests is less (e.g., in residential areas). Further critiques [11] of the Channel Reservation approach are that it is unable to ensure fair Quality of Service (QoS) to different types of services (e.g., more bandwidth for wide-band handover requests) and is inefficient under varying traffic conditions.

The other classical technique, *queuing of handover requests*, allows requests to queue in First in First out (FIFO) order [12]. Priority queuing includes Pre-emptive Resume (PR) and Head of Line (HoL) among others [28]. The queuing of requests is possible due to the overlapping of cells, called the handover area. Requests remain in the queue until channels become available or the signal of the requests drops to a very low level. In the latter situation, the requests are blocked [13]. As in the above example, if a channel is available and the user is still in the handover area, the channel is allocated to her request with the highest priority, depending on the type of queuing technique used. Handover requests are guaranteed to get high priorities as new requests are not processed until there is space in the queue. The advantage of this scheme is that it reduces the probability of forced termination of handover requests. The disadvantage is that it increases request blocking probability [14]. But both of these quantities determine the QoS of cellular systems. Some researchers [15] claim that the queuing scheme with FIFO policy exhibits performance very close to that of the ideal prioritized handover scheme, while others [16] disagree with this claim, arguing that this mechanism does not take into account the dynamics of user motion.

Guerin's has developed a method, called Guard Channel Method [18], in order to prioritise handover requests. This method combines channel reservation [19] with queuing mechanism [20]. It reserves a number of channels for handover requests and allows new requests to buffer in an infinite queue. In contrast to the conventional handover, queuing of new requests minimises the blocking of fresh requests while maintaining low probability of handover blocking by reserving channels. Hong et al [9] tested a similar approach to Guerin's method but they only permitted handover requests to buffer in infinite queue with FIFO policy. Infinite queue may not represent a realistic system. Chang et al [21] proposes another approach to prioritise handover requests in which handover requests and fresh requests are queued separately in finite queues. In addition, it enables the queued new requests to renege and queued handover requests to drop if they move outside the handover area before the handover request is granted. The authors believe that finite queuing mimics the real world better than infinite queuing although implementing two separate queues may be a costly approach. Jabbari et al [22] consider the movement of mobile users with different speeds. They promote a non-preemptive dynamic priority queuing discipline for handover requests called the measurement based priority scheme (MBPS). This scheme is based on power measurements. The rate of degradation of the power levels of the queued requests are measured continuously. When a channel becomes free, the request with the lowest signal level is granted the service. The simulation results found in [22] indicate that the proposed scheme offers a better QoS with lower handover blocking probability, spectrum utilization and less delay. Xhafa et al [16] underline that the scheme does not consider the dynamics of user mobility as mobile users move with different speeds.

In the next section, we present a framework for the performance evaluation of cellular networks. The analysis of this framework is based on decomposition technique and non-preemptive priority queuing scheme, called Head of Line (HoL) — which is enhanced through a cut-off priority technique using queue threshold.

3 The Proposed Approach

One of the most popular wireless network architectures is known as a cellular network. According to this architecture, the geographical area is divided into a number of regions, formally known as coverage areas or cells (c.f., Figure 1).

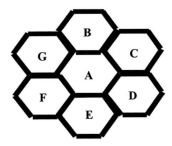

Figure 1: The seven-cell cluster

Communications in each cell are controlled by a base station. These base stations are connected to wide area networks such as *public switched telephone networks* (PSTN) and Internet as shown in Figure 2.

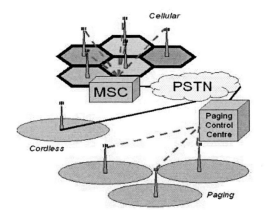

Figure 2: Cellular network architecture

Users communicate or access various services using their mobile phones or PDAs. Each such device at all the times throughout its communication session remains attached to the base station. Each cell supports mobile devices using scarce available channels. As soon as a user moves from one cell to another, the point of attachment is changed as well in order to maintain the ongoing connection. Providing a seamless connectivity to roaming users poses various challenges to research community (c.f., Figure 3). Effective mechanisms for mobility management must be developed in order to improve the quality of service.

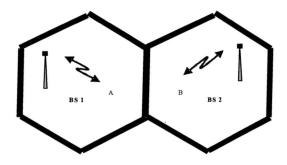

Figure 3: Handover scenario at cell boundary

A cellular network can be represented as a queuing network model (QNM) where each cell is a queuing node with finite capacity buffer (queue) to temporarily hold the incoming requests, and a number of available channels for providing connectivity. Each cell deals with two types of requests, namely, *handover requests* and *new requests*. In the subsequent discussion we use HO to represent *handover requests* and NR to represent *new requests*. Such requests are made to access a set of available channels. In order to efficiently manage such requests and to improve the overall performance of the system, it is required to devise an effective buffer management scheme and to ensure an efficient use of scarce channels. Accordingly, this work proposes a framework for performance evaluation of cellular networks. This framework extends our previous analysis, focusing on a single cell (c.f., [30]), to multiple cells considering the blocking of requests (or calls) at the time of handover. The proposed framework is based on an effective buffer management scheme with queue thresholds and service priority mechanism.

According to the proposed framework, a cellular network is decomposed into individual cells. Each cell is then analysed as a GE/GE/C/N/HoL queuing system. This model is believed to provide a mechanism for an effective use of available channels by HO as well as NR requests. Inter-arrival and service times at each channel for both types of requests are represented by GE distribution (c.f., Figure. 4). C represents the total number of channels available in the cell. Vector $N = (N_1, N_2)$ represents the queue thresholds with N_2 as the threshold and N_1 as the total capacity of the queue to temporarily hold the incoming NR and HO requests. The proposed scheme processes NR and HO requests in a following manner:

- Whenever a system receives a request of any type, (NR or HO), it is accommodated in the queue provided the queue length has not reached the threshold value.
- If the queue length reaches the threshold value, then it only allows HO requests to enter the queue. In this case, it blocks all the NR requests.

Free channels are assigned to the requests in the queue according to HoL scheduling discipline. This adds to the unique features of the proposed framework from the existing ones. Restricted number of requests capacity in the queue mimics the limited number of channels available due to scarce frequency spectrum availability. Such an approach has the advantage of improving the perceived QoS since fewer high priority requests (HO) will lose the connection due to the additional protection introduced by the space priority. For example, during the NR and HO requests, if all the channels are occupied, then both types of requests are allowed to queue. If the queue already holds HO and NR requests, and a new HO request arrives at the queue, then it is placed before all the queued NR requests following the HoL queuing discipline but remains at the end amongst the HO requests, the secondary queuing policy among HO requests being FIFO. This gives HO requests more priority than NR requests. Permitting NR requests to queue does not block the requests if no channels are free but are simply delayed and will ultimately receive the service, contributing to the increasing carried traffic. No requests are pre-empted once the channels are allocated. When a channel becomes free upon completion of a request, it is assigned to the request at the head of the queue.

Traffic Models

As discussed above, one of the key parameters that evaluate the performance of the system is the HO and NR loss or blocking probability. In order to satisfy the blocking probability requirement, HO and NR traffic rates have to be estimated accurately under realistic assumptions. Firstly, we consider that the cellular traffic is bursty in nature — meaning, requests tend to arrive in bulks. Poisson distribution (c.f., [24, 25]) is the most commonly used distribution to model the number of events occurring within a given time interval. Until when Chlebus et al [26] first questioned if handover traffic was always Poissonian, most of the models assumed to have Poissonian traffic to model the number of events occurring within a given time interval. A Poisson arrival process does not adequately characterise arrival traffic for cellular networks as the theory assumes one arrival at random time. It is only applicable in a non-blocking environment as in telephone networks, but not in growing internet traffic. In cellular networks, there exists request blockage during the unavailability of sufficient resources. As described above, we consider that the cellular traffic is bursty in nature. The following section briefly describes the Generalised Exponential (GE) Distribution [27, 28] which is used to model the traffic with burstiness property.

The GE Distribution

The GE distribution is a mixed interevent-time distribution of the form [27]:

$$F(t) = P(X \leq t) = 1 - \tau e^{-\sigma t}, t \geq 0,$$

where $\tau = 2/(C^2 + 1)$, $\sigma = \tau \upsilon$ and X is the interevent time random variable (rv). $\{1/\upsilon, C^2\}$ are the mean and the Squared Coefficient of Variation (SCV), the ratio of the variance to the square of the mean, of the interevent time distribution X, respectively. Raad et al [29] state that all distributions can have a SCV higher than 1 except the exponential distribution which has a coefficient of 1. SCV provides an important measure of the variability of the distribution [27].

The counting process of the GE distribution is a Compound Poisson Process (CPP) with parameter $2\upsilon/(C^2 +1)$ and a geometrically distributed batch sizes with mean $1/\tau$, $(C^2 +1)/2$ and SCV, $(C^2 -1)/(C^2 +1)$. GE distribution is useful for modelling random variables with SCV greater than 1.

The GE distribution has a memory-less property [27] like Poisson distribution. The GE distribution is considered to be the most appropriate distribution to model the inter-arrival times of HO and NR requests as it supports simultaneous arrivals, unlike Poisson distribution. In this context, the burstiness of the arrival process is characterised by the SCV of the inter-arrival time [27].

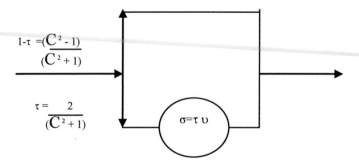

Figure 4. The GE distribution with parameters τ and σ $(0 \leq \tau \leq 1)$

4 GE/GE/C/N/HoL Queueing Model Description

This section describes the GE/GE/C/N/HoL queueing system (c.f., [30]) which represents each cell in a cellular network (c.f., Figure 5).

Let, i represents the classes of requests (HO and RO requests), λ_i and μ_i represent the mean arrival rate and service rate, C_{ai}^2 and C_{si}^2 are the SCVs of the inter-arrival and service times for the ith class, respectively. We consider that the values of μ_i and C_{si}^2 for all types of requests are the same as we assume that all channels have the same service rates.

When a GE arrival process with rate λ_i is sampled with probability $\tau = 2/(C_{ai}^2 + 1)$, the GE arrival rate ($\sigma = \tau \lambda_i$) of the request will be $2\lambda_i /(C_{ai}^2 + 1)$, and the GE inter-arrival time will be $(C_{ai}^2 + 1)/ 2\lambda$.

Likewise, for the GE service process with rate μ_i, the probability that the request will receive service is $2\mu_i/(C_{si}^2 + 1)$ and, the GE inter-service time is $(C_{si}^2 + 1)/ 2 \mu_i$

5 Numerical Results

This section presents the experimental results and evaluates the performance of the proposed framework. The performance of the proposed framework for a cellular network architecture has been evaluated using QNAP-2 [17] simulation package. Each cell in the cellular network is represented by a GE/GE/C/N/HoL queue with two types of arriving requests (i.e., HO and NR). Various experiments have been conducted on a two cell model connected to each other (c.f., Figure 6). HO requests follow repetitive service blocking mechanism at the time of transfer from one base station to another base station. The outputs of the experiments performed on the proposed enhanced priority queuing framework with HoL policy and buffer threshold are captured and analysed.

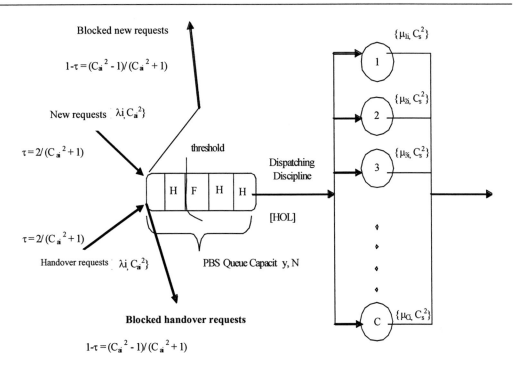

Figure.5.The proposed Priority Queuing Handover Scheme with HoL policy and threshold in finite Buffer

The performance evaluation of the proposed framework is conducted at two levels. The first level models the number of channels in the cell as the varying input parameter and the second level models the handover mean arrival rate as a varying input parameter (c.f., Table 1). Both levels determine the performance of the proposed framework on the basis of response time, and loss or blocking probability for HO and NR requests.

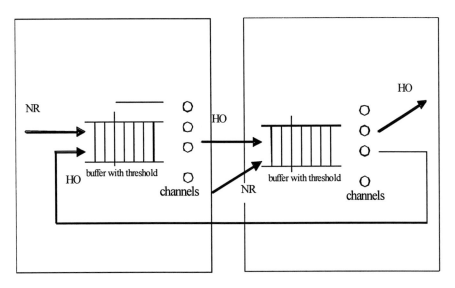

Figure 6: QNM representing two cells architecture

The following input parameters have been used in all simulation experiments. Each experiment is subdivided into two sets:

SET A: Number of channels is varied from 1- 8 at both cells
SET B: Handover traffic load is varied from 1.0 - 4.5 at both cells

Generally it is assumed that most of the traffic generated within a cell is of type NR requests. Hence, the HO arrival rate in each set of experiments is less than the NR arrival.

All the channels provide the same service to either of the request types. Therefore, the service rate for both types of requests is kept the same. SCV for both types of requests are put higher than 1 to model the burstiness of the traffic. To minimize system congestion, arrival rate of both types of requests are left lower than the service rate. Experimental results are explained as follows:

Experiment 1: Figure 7 shows the results from the experiment when HO and NR perform the priority queuing scheme with HoL policy and buffer threshold in increasing number of channels. The graph highlights the loss probability of HO and NR requests. Initially, the loss probability for HO requests is much less than that of the NR requests but the loss probability of high priority (HO) and low priority (NR) requests decreases with the increase in the number of available channels. The high loss probabilities of NR and low loss probability of HO occur when there is a minimum channel resource which represents the system performance during scarce spectrum resource.

TABLE 1: Simulation Experiments: System Input Parameters

System parameters	Set A – Variation in number of channels	Set B- Variation in handover traffic load
Number of Channels	1 – 8	3
Queue Capacity	20	20
Threshold value	15	15
Handover Arrival rate	3	1.0– 4.5
SCV for handover arrival requests	3	3.0
Handover Service rate	8.0	8.0
SCV for handover service rate	3.0	3.0
New requests arrival rate	7.0	7.0
SCV for new arrival requests	3.0	3.0
New Requests service rate	8.0	8.0
SCV for new requests service rate	3.0	3.0

Experiment 2: This experiment on priority queuing scheme with HoL policy and buffer threshold (as shown in Figure 8) demonstrates the impact of HO traffic load on the loss

probabilities for HO and NR requests. It is understood from the graph that the loss probability for HO and NR requests increases with the increase in the HO traffic load. It can be seen that for higher traffic load the difference between the loss probabilities of NR and HO requests increases. The loss probability for HO requests is much less than that of new requests.

Experiment 3: This experiment is conducted to represent the system response time for both types of requests against various number of channels. As shown in Figure 9, the delay for NR is comparatively higher than that of HO for smaller number of available channels. Response time for NR and HO requests decreases by increasing the number of channels.

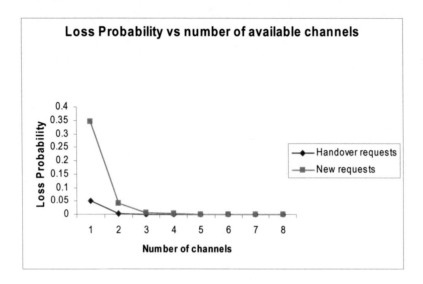

Figure 7. Loss probability of HO and NR in priority queuing handover with buffer threshold

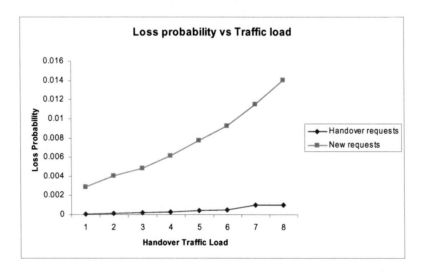

Figure 8. Loss probability of HO and NR in priority queuing handover with buffer threshold

Experiment 4: This experiment has been carried out to test the behaviour of the system's response for higher number of HO requests. The graph (c.f., Figure 10) shows a gradual increment of response time for HO and NR when the HO traffic load increases. The response time for HO requests is lower than that of NR. The increasing traffic load does not make significant impact on response time of HO and NR unlike in the previous experiment.

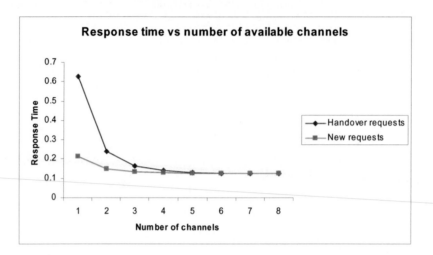

Figure 9. Mean Response Time of HO and NR in priority queuing handover with buffer threshold

The above experiments validate that the proposed priority queuing handover mechanism in finite queue provides services with lower loss probability and lower mean response time of HO than that of NR. This means the scheme takes into account that HO requests are of higher priority. Thus their blockage can result in poor service. As previously discussed, the QoS for the proposed framework is defined as low handover blocking probability and minimized new request blocking probability. In both instances (Experiments 1-2), the loss probability of HO is less than that of NR. Similarly, response time for NR requests is higher than the HO requests in the proposed system. This simulation model proves that the proposed framework minimizes the trade-off of handover blocking probability with high request blocking probability.

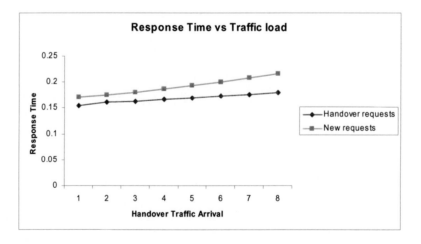

Figure 10. Mean Response Time of HO and NR in priority queuing handover with buffer threshold

6   Conclusion and Future Prospects

This chapter proposes a framework for the performance evaluation of mobility management in cellular networks. This framework has been modelled by a queuing network system where each cell is represented by a GE/GE/C/N/HoL queue with threshold and HoL service priority. It is based on decomposition technique such that cells are evaluated in isolation. The performance of the system has been demonstrated using a two stations queuing system to see the behaviour of mobility management under the proposed threshold based queuing management and service priority.

The proposed framework was rigorously evaluated through a set of experiments. Experimental results demonstrate that the proposed framework:

- ensures high reliability by reducing the loss probability of HO requests during the availability of limited channels. Though it results in higher loss probability for NR requests, it occurs only when there is a minimum channel resource which represents the system performance during scarce spectrum resource.
- ensures high reliability by reducing the loss probability of HO requests during higher traffic load. It shows that with the increase in traffic load the difference between the loss probabilities of NR and HO requests also increases.
- shows that the delay for NR is comparatively higher than that of HO for fewer available channels. By increasing the number of channels, the response time for these types of requests reduces. This shows the effectiveness of the proposed priority based scheme when resources are very limited.
- demonstrates less response time for HO requests as compared to NR requests for increasing traffic load.

In summary, the experimental study demonstrates that the enhanced priority queuing scheme with buffer threshold provides QoS with low connection failure and mean response time of handover requests. Our future research will investigate the effect of channel allocation schemes to various cells for minimizing the channel interference on mobility management algorithms.

References

[1]   M. Mohapatra, and V. Pachaury, "A fixed wireless cellular alternative to wireline telephony", IEEE International Conference on Personal Wireless Communications, 1994, Aug. 1994 (Bangalore, India) pp 64 -72.

[2]   R. Ramjee, R. Nagarajan, and D. Towsley, "On optimal call admission control in cellular networks", The IEEE INFOCOM 96: Fifteenth Annual Joint Conference of the IEEE Computer and Communications Societies, March 1996 (San Francisco, USA) pp 43-50.

[3]   Unknown Author(Cellular.co.za), "How a GSM network Operates", [Available: http://www.cellular.co.za/howagsm.htm, Accessed: October 13, 2004, April 12, 2005]

[4]   J. Li, N.B. Shoff, and E.K.P Chong, "Channel carrying: A novel handoff scheme for mobile cellular networks", *IEEE/ACM Transactions on Networking*, Vol. 7, No. 1, pp 38-50 (1999).

[5]   P. Marichamy, S. Chakrabarti, and S.L.Maskara, "Overview of handoff schemes in cellular mobile networks and their comparative performance evaluation" Vehicular Technology Conference (VTC 99), Sept 1999 (Amsterdam, Netherlands) pp 1486–1490.

[6]   Y-B Lin, A.R. Noerpel, and D.J. Harasty, "The sub-rating channel assignment strategy for PCS hand-offs", *IEEE Transactions on Vehicular Technology*, Vol. 45, No. 1, pp 122–130 (1996)

[7]   D.K. Anvekar, and S.S Pradhan "HCE: A new channel exchange scheme for handovers in mobile cellular systems", IEEE International Conference on Personal Wireless Communications, Feb 1996 (New Delhi, India), pp 129-133

[8]   M. Wu, W.E Wong, and J.J Li, "Performance evaluation of predictive handoff scheme with channel borrowing", Performance, Computing, and Communications Conference, April 2003, pp 531–536

[9]   D. Hong, and S. S Rappaport, "Traffic model and performance analysis for cellular mobile radio telephone systems with prioritised and nonprioritized handoff procedures", *IEEE Transactions on Vehicular Technology*, Vol. 35, No. 3, pp77-92 (1986)

[10] W. Zhuang, B. Bensaou, and K.C Chua, "Adaptive quality of service handoff priority scheme for mobile multimedia networks", *IEEE Transactions on Vehicular Technology*, Vol. 49, pp 494-505 (2000).

[11] S.S. Rappaport, and C. Purzynski, "Prioritized resource assignment for mobile cellular communication systems with mixed services and platform types", *IEEE Transactions on Vehicular Technology*, Vol.45, No.3, pp 443-457 (1996).

[12] Unknown Author (Unknown Date), "Cellular Communication", Web ProForum Tutorials- The International Engineering Consortium (burnsidetelecom.com), [Available: http://www.burnsidetelecom.com/whitepapers/cell_comm.pdf, Accessed: April 26, 2005]

[13] J.D. Gibson, "The mobile communications handbook", *CRC/IEEE Press* (1996).

[14] S. Tekinay and B. Jabbari, "An effective prioritization scheme for handovers in cellular networks", First International Conference on Universal Personal Communications, Sept. 1992 (Dallas, Texas)

[15] R. Fantacci, "Performance evaluation of prioritized handoff schemes in mobile cellular networks", *IEEE Transactions on Vehicular Technology*, Vol. 49, Issue 2, pp 485–493 (2000)

[16] A.E. Xhafa, and O.K. Tonguz, "Dynamic priority queuing of handover calls in wireless networks: An analytical framework", *IEEE Journal on Selected Areas in Communication*, Vol. 22, No. 5, pp 904-916 (2004)

[17] M. Veran, D. Potier, "QNAP -2: a portable environment for queuing systems modelling" in D. Potier (Ed.), Modelling Techniques and Tools for Performance Analysis, North HoLland, Amsterdam, 1985

[18] R. Guerin, "Queueing-blocking system with two arrival streams and guard channels", *IEEE Transactions on Communications*, Vol. 36, No. 2, pp 153-163 (1988)

[19] E.C Posner, and R. Gueri, "Traffic policies in cellular radio that minimize blocking of handoff calls", ITC-11, Kyoto (1985)

[20] Unknown Author (2004), "Cellular communication-digital systems", IEC: On Line Education(IEC.org), [Available: http://www.iec.org/online/tutorials/cell_comm/ topic06.html?Next.x=40&Next.y=8, Accessed: April 20, 2005]

[21] C-J Chang, T-T Su, and Y-Y Chiang, "Analysis of a cutoff priority cellular radio system with finite queueing and reneging/dropping", *IEEE/ACM Transactions on Networking,* Vol. 2, No. 2, pp 166 -175 (1994)

[22] S. Tekinay, and B. Jabbari, "A measurement-based prioritization scheme for handovers in mobile cellular networks", *IEEE J. Select. Areas Communication*, Vol.10, pp 1343-1350 (1992)

[23] H. Schulte, and w. Cornell, "Multi-area mobile telephone system", *IRE Transactions on Vehicular Communications,* Vol. 9, No. 1, pp 49–53 (1960)

[24] H.G. Ebersman, and O.K. Tonguz, "Handoff ordering using signal prediction priority queuing in personal communications systems", Sixth IEEE International Symposium on Personal, Indoor and Mobile Radio Communications (PIMRC'95) Sept. 1995, pp 824-828

[25] A.H Ahmed, "Characterization of beta, binomial, and Poisson distributions", *IEEE Transactions on Reliability*, Vol. 40, No. 3, pp 290-295 (1991)

[26] E. Chlebus, and W. Ludwin, "Is handoff traffic really Poissonian?", Fourth IEEE International Conference on Universal Personal Communications, Nov. 1995, pp 348 -353.

[27] D.D Kouvatsos, "Entropy maximization and queuing network models", *Ann. Oper. Res.* Vol. 48, pp. 63-126 (1994)

[28] I. Awan, and D.D Kouvatsos "Entropy maximization and open queuing networks with priorities and blocking", Special Issue on Queueing Networks with Blocking, Performance Evaluation, Department of Computing, University of Bradford, UK, Performance Evaluation-Vol. 51, Issue: 2-4, pp 191- 227 (2003)

[29] R.S. Raad, and E. Dutkiewicz, and J. Chicharo, "Connection admission control in micro-cellular multi-service mobile networks", Fifth IEEE Symposium on Computers and Communications, July 2000, pp 600 – 606

[30] I. Awan and S. Singh, "Performance evaluation of E-commerce requests in wireless cellular networks" *Information and Software Technology,* Vol.8, No.6, pp 393-401 (2006)

In: Mobile Multimedia: Communication Engineering ...
Editors: I.K. Ibrahim and D. Taniar pp. 17-42

ISBN: 1-60021-207-7
© 2006 Nova Science Publishers, Inc.

Chapter 2

PERSONALISATION AND CONTINUITY OF MOBILE SERVICES

*Ivar Jørstad**

Department of Telematics, Norwegian University of Science and Technology, O.S.
Bragstads Plass 2E, NO-7491 Trondheim, Norway

*Do van Thanh***

Telenor R&D, Snarøyveien 30, NO-1131 Fornebu, Norway

Abstract

This chapter focuses on two challenging issues in mobile services, namely Service Continuity and Service Personalisation. While Terminal mobility refers to the ability of the terminal to change locations without loosing connection, service continuity designates the ability of services to be uninterrupted when the terminal is moving. Service continuity has two facets. One facet is on the perception of the user that experiences the service as continuous. The other facet concentrates on the ability of the service of surviving the change of cells, networks and service components. Services can be synchronous or asynchronous, real-time or non-real time and will have different requirements. Only a few current services do really support service continuity. In this chapter, a new architecture for mobile services will be introduced and explained thoroughly. The most important characteristic is the dynamic composition aspect of a mobile service which is realised by a dynamic assemblage of service components. The mechanisms and functions necessary to enable service continuity will also be described in details.

Human beings are born different and will have different requirements on mobile services. Service personalisation allows the users to customise services as they please. After that, no matter where they are, which device or networks they are using, the services will have the same look-and-feel and behaviour as the user expects. Service personalisation calls also for a new architecture where the user profile that captures the user preferences and settings can be distributed and maintained by different players. This chapter will describe the mechanisms and functions that are necessary to realise service personalisation. Last but least, the chapter will also treat the complication that arises when both service continuity and service personalisation must be enabled together.

---

* E-mail address: ivar@ongx.org
** E-mail address: thanh-van.do@telenor.com

Introduction

The goal of this chapter is to give the reader an in-depth understanding of two important concepts of future mobile services, namely *Service Continuity* and *Service Personalisation*. Until now, these topics have not been thoroughly investigated, although both of them are universally considered to be important qualities of mobile services, which can increase the user adoption rate of new services.

In mobile telecommunication it is common to use the term "terminal mobility" to denote the ability of a user with a handset to move geographically while still receiving the same service (i.e., mobile voice telephony). Terminal mobility is in for example GSM solved by a combination of handover and roaming. Handover is used to switch between base stations when the signal level gets to low on the current one, whereas roaming is used to let a user keep the same service when moving outside the coverage of the home network (e.g. moving abroad). In GSM, these procedures are tailored for the voice telephony service, since this was the primary service to be delivered by these networks. However, more and more services are now accessed through wireless and mobile devices, connected to different networks using different technologies. Since the handover procedures in GSM are developed for the voice service, for other services it only means that a physical connection will be available. There exists no solution that takes into consideration all the other elements of services that are of importance to maintain the service when the user moves between devices, networks and service domains.

Service Continuity denotes the ability to maintain service access when a user moves between devices, networks or service domains. Like handover is realised on the physical and network layers, service continuity is a function located in the middleware right below the application layer. The goal of this function is to offer for any type of service, the same as handover and roaming does for voice telephony in GSM. However, connectivity is assumed to be in place (services still run over the same phyiscal bearers). It will be shown in this chapter that there are several obstacles preventing the realisation of the service continuity function, and possible solutions to these will be discussed.

Closely related to Service Continuity, but also with its own challenges, is the concept of Service Personalisation. Personalising services is to adapt services to fit the needs and preferences of a user or a group of users.

Personalisation is important in today's service-oriented society, and has proven to be crucial for the acceptance of services provided by the Internet and mobile telecommunication networks (illustrated by the success of personalised ring tones, logos, etc.). In [1], the motivation for personalisation is described and two important categories of personalisation are identified: *personalisation to facilitate work* and *personalisation to accommodate social requirements*.

In the first category, services are adapted to increase the efficiency, e.g. to minimize the time spent on repetitive and similar work tasks. The adaptation can aim at accommodating physical differences of the users like weak sightedness, disabilities, etc.

In the second category, services are adapted to enhance the social experience. For example, youngsters, by changing the appearance and behavior of a cellular phone (ring tone, logos etc.) want to express their identity/personality.

To enable service developers and providers of both the Internet and mobile telecommunication networks to support personalisation, adequate middleware, service platforms and environments must be present. They act as a fundament and catalyst for increasing the number of personalised services.

The first sections of this chapter will analyse existing mobile services architectures to illustrate their characteristics, strengths and weaknesses. The next sections will discuss service continuity and service personalisation. Thereafter, the challenges of the two are considered together, since they must both be supported by the same service execution environment.

Analysis of Current Mobile Service Architectures

Until now, mobile telephony and Short Message Service (SMS) are the only successful mobile services. This section will analyse the architectures of these two services.

Mobile Telephony

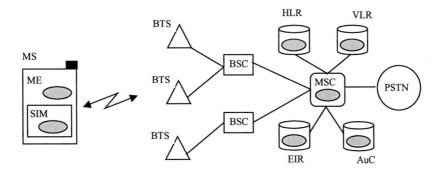

MS: Mobile Station – ME: Mobile Equipment – SIM: Subscriber Identity Module
BTS: Base Station Transceiver – BSC: Base Station Controller – HLR: Home Location Register
– MSC: Mobile service Switching Center - VLR: Visitor Location Register – EIR: Equipment
Identity Register - AuC: Authentication Center – PSTN: Public Switched Telephony Network

Figure 1 Mobile telephony service architecture

The architecture of the mobile telephony service in GSM (Global System for Mobile communications) [2] is depicted in Figure 1. The mobile terminal or Mobile Station (MS) in GSM terminology consists of two separate devices: the Mobile Equipment (ME), which is the mobile phone itself and the Subscriber Identity Module (SIM), which is a smart card containing the identity of the subscriber and the authentication keys and functions. When the mobile terminal is switched on, the service logic component (shown in grey ovale in Figure 1) on the ME in the collaboration with the one of the SIM carry out the authentication towards the Visiting Location Register (VLR), Home Location Register (HLR) and Authentication Center (AUC). Upon success, the mobile terminal is ready to initiate or receive calls. A phone call is established through interactions between service logic components located on the mobile terminals and those located on the Mobile Switching Center (MSC), VLR and HLR.

When the mobile terminal is moving, the local service logic components will interact with the service logic components located on other MSCs and VLRs. These service logic components can be implemented by different telecommunication manufacturers but they all have the same functionality and interfaces that are rigorously specified by the European Telecommunications Standards Institute (ETSI).

The mobile telephony service is realised by static service logic components that are installed in the mobile terminal in collaboration with static components installed on the mobile networks. When the mobile terminal is moving, the continuity of the mobile telephony service is ensured by the fact that the service logic components on the terminal are interacting with *different equivalent service logic components on the mobile network.* Although there is a plethora of mobile terminals in the market, they are made up of equivalent service logic components built according to well-defined standards.

Briefly, it is possible to conclude that:

The mobile telephony service is realised by a dynamic composition of static equivalent service logic components.

Further, the mobile telephony service is only provided in the home country and the countries where there is a roaming agreement between the visited operator and the home operator. A roaming agreement between two or more mobile operators outlines the terms and conditions under which the participating operators will provide mobile services to each others subscribers.

It is possible to conclude that:

The dynamic composition is pre-defined and only "trusted" static equivalent service logic components are selected.

It is also worth noting that the formation of the mobile telephony is decided by the networked service logic components after successful authentication of the mobile terminal. In 3G networks, both the mobile terminal and the network have the right to decide to engage in mobile telephony after a mutual authentication.

Short Message Service (SMS)

The Short Message Service architecture [3] is depicted in Figure 2. The service is realised roughly by three service logic components: one in the mobile phone, one in the SIM card and one in the SMS-C server. Independently of the movement and location of the user and the mobile phone, the same logic components will be used.

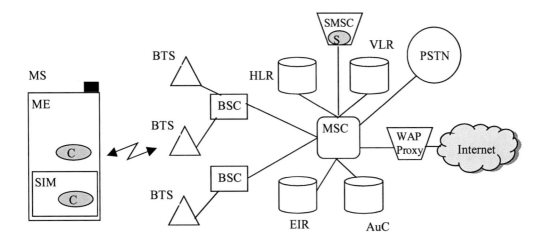

Figure 2 Short Message Service architecture

It is possible to conclude that:

The Short Message Service is realised by a static composition of static identical service logic components.

Requirements on Next Generation Mobile Service Architecture

This section provides an informal treatment of the high-level requirements of the next generation mobile service architecture, where the goal is to illustrate some of the challenges. For the next generation mobile service there will be many more requirements that the architectures of mobile telephony and SMS cannot fulfil. The main high level requirements are as follows:

A. The architecture must be flexible and extensible to cover heterogeneous services.

The next generation mobile services will be of different types: asynchronous versus synchronous, real-time versus non real-time, communication versus computing, single user versus multiple users.

B. The architecture must be flexible and extensible to support heterogeneous devices

The user now has several devices which are similar, but not necessarily completely equal in characteristics (e.g. cellular phone, PDA and laptop). Some of these can deliver similar types of services to the user.

C. The architecture must accommodate several distribution schemes: monolithic versus multiple components, client-server versus peer-to-peer, static code versus mobile code.

D. The architecture must allow a service to be composed dynamically by equivalent service logic component

E. The architecture must allow dynamic service composition of both pre-defined and ad-hoc service logic components.

Indeed, it is very important to be able to realise a service by dynamically discovering, selecting, assembling and executing the service logic components.

It is also crucial, with the user's move and location change, to be able to dynamically detect and use new equivalent service logic components instead of the former ones.

It is also necessary that the same service logic components can move and change locations.

The architecture of a mobile service is shown in Figure 3. A service can be realised by a multitude of assemblage variants.

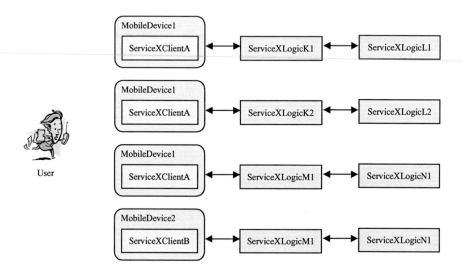

Figure 3 Architecture of a next generation mobile service

In the first alternative, when the user is using a *MobileDevice1*, a *ServiceX* is realised by *ServiceXClientA*, *ServiceXLogicK1* and *ServiceXLogicL1*.

When the user is moving but using the same mobile device, the *ServiceX* is realised by the same *ServiceXClientA* but the logic components *ServiceXLogicK1* and *ServiceXLogicL1* are replaced respectively by new instances *ServiceXLogicK2* and *ServiceXLogicL2* as shown in alternative 2.

In alternative 3, the logic components *ServiceXLogicK1* and *ServiceXLogicL1* are replaced respectively by equivalent implementations *ServiceXLogicM1* and *ServiceXLogicN2*.

When the user changes to another *MobileDevice2*, the *ServiceX* is realised by a new *ServiceXClientB* and the same *ServiceXLogicK1* and *ServiceXLogicL1*.

It is worth noting that there is nothing special about a service realised by several distributed components. The important point is that a mobile service is realised by components that are replaced dynamically by other equivalent components. This knowledge is used in the next sections to properly analyse the requirement of *Service Continuity*.

Service Continuity

A necessary requirement for any mobile service is *service continuity* [4] i.e. a service does continue to operate without interruption when the user is moving, either together with the mobile device, or between mobile devices. In mobile communication, the terms terminal mobility denotes the continuity of telephony service which is tightly related to the handover at the network level. The service continuity is more general and applies for all services. The service continuity relies on three elements:

1. The network connection: without network connection, a mobile service will definitely be discontinued. Network handover is still a necessary condition
2. The ability of the mobile service to survive, adapt to changes and continue execution.
3. The user experience of continuity: depending on the type of the service, continuity is experienced differently by the user.

In this section, only the second element will be considered since it is directly related to the architecture of the mobile service.

When the user and device are moving to a new location, before loosing communication with the former ServiceXLogicK1, it is necessary that the ServiceXClientA discovers and uses the ServiceXLogicK2.

To discover a service logic component means:

1. To detect the available service logic components in the current location: This can be done by broadcasting and waiting for answer from the local service logic components. However, in order to reduce the detection time and ensure continuity, it may be necessary to have *Registry Component* that keeps track of all available service logic components and reports to the requesting client.
2. To verify that both the semantic and the syntax of the detected service logic component matches with the former one. As stated in [5], a service logic component is equivalent with another one if both their semantics and syntaxes are equivalent.
3. To initialize the new service logic component to continue from where the former has arrived.

Consider now the transition when the service logic components *ServiceXLogicK1* and *ServiceXLogicL1* are replaced respectively by new instances *ServiceXLogicK2* and *ServiceXLogicL2*.

To achieve service continuity, the following conditions must be fulfilled:

o   Before abandoning *ServiceXLogicK1* and *ServiceXLogicL1,* the *ServiceXClientA* must detect, recognise, authenticate and establish communication with the new instances *ServiceXLogicK2* and *ServiceXLogicL2*.

o To be able to continue where *ServiceXLogicK1* and *ServiceXLogicL1* left, the *ServiceXLogicK2* and *ServiceXLogicL2* need to be supplied with internal state data of *ServiceXLogicK1* and *ServiceXLogicL1*.
o The internal state of *ServiceXLogicK1* and *ServiceXLogicL1* or more generally, the internal state of the executing ServiceLogic components must be captured and made available to the next replacing components.

Consequently, as stated in [4], the ServiceData should be separated from the ServiceLogic. The architecture of a mobile service is shown in Figure 4.

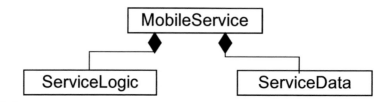

Figure 4 Separation of Service Logic and Service Data

In fact, to ensure continuity when the user and the mobile device is moving and *ServiceXLogicK1* is replaced by *ServiceXLogicK2*, the internal state of *ServiceXLogicK1* must be recorded by *ServiceXDataK1* which is made available to ServiceXDataK2 which is the internal state *ServiceXLogicK2*. *ServiceXLogicK2* is hence able to resume execution where ServiceXLogicK1 suspended such that *ServiceX* can appear continuous to the user.

It is hence required that both *ServiceXLogicK1* and *ServiceXLogicK2* have the same internal variables which values can be exported to or imported from a global *ServiceXData* component which is known by both as shown in Figure 5a.

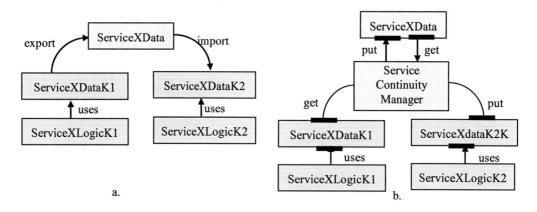

Figure 5 Service Continuity Manager

Unfortunately, services are not structured with a well-defined ServiceData and the ServiceLogic is not equipped with interfaces to export or import the values of ServiceData. A solution is to have a Service Continuity Manager that extracts and saves the service data values in ServiceData in order to supply them to the next executing ServiceLogic as shown in Figure 5b.

Service Content

Most services are involved with some contents. The contents can be of different types as follows:
- Voice, video, data
- Structured, unstructured
- Volatile, persistent

For example, the content of telephony is voice which is unstructured and volatile. This content becomes persistent when it is recorded. For a word processor, spreadsheet application, or presentation program, the content i.e. document, spreadsheet or presentation is very central.

To preserve continuity, when the user and device are moving, it is necessary to capture the content with latest changes and transfer it to the next executing service logic component. It is therefore reasonable to separate the ServiceContent from the ServiceLogic as shown in Figure 6 and to make use of the Service Continuity Layer [6] middleware to ensure the continuity of the ServiceContent.

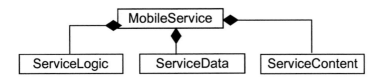

Figure 6 Separation of Service Content from Service Logic

Service Personalisation

Another requirement on the next generation mobile services is personalisation. Several different definitions of personalisation can be found but in this chapter the definition derived from [7] is used:

"Personalisation of a service is the ability to allow a user U to modify or produce, a service A such that it fits user U's particular needs in terms of presentation and functionality, and after such personalisation, all subsequent service rendering of service A for user U will be conformed to the performed modification.."

Although the definition states that personalisation is done by the user, most of the tasks are done by either the service itself or the service platform. In fact, it only stresses that the user should be the one in charge and initiating the personalisation process in the first place.

From the definition it is possible to deduce the first requirement:

The service itself must be built in such a way that it is modifiable in execution time.

This requirement may seem to be trivial but it is not. Indeed, at early stage of computing, software programs on mainframe were quite rigid and it was mostly impossible to alter the functionality at run time. Any change had to be done at code level and recompilation was required. The presentation was even less flexible since no graphical but only text interfaces were available. When using the software program, the users had to adapt themselves to it.

A service that is modifiable is commonly called **customisable**. **Customisation** *refers to the functionality and presentation modification to suit a customer i.e. the one who bought or subscribed the service.* It does not take into the different persons that use the service. Indeed, all the users of the same subscription will experience the service identically. If one user applies a change to the service, this change will be exposed to all the users.

A service is customisable if the parameters/variables for presentation and functionalities can get assigned values from a defined set of values.

It is also possible to deduce the following corollary:

If a service is personalisable it is customisable.

The inverse is not true since not all customisable services are personalisable. In order to extend a customisable service to be personalisable, each user should get allocated a distinct personalisation information set. This personalisation information set is often called User profile [8][9] or User preferences [10], but the term personalisation information set is more general (it covers additional information compared to a traditional user profile, and certainly more than user preferences), and thus more appropriate. Every time a user starts the service his personalisation information set shall be used to adapt the service. It is hence necessary to *recognize the user and to make sure that he is the right one.*

The second requirement is therefore:

To enable personalisation, the identification and authentication of the user must be provided.

If the user always uses the same service installed in his PC or accesses the same service instance installed in one server from one device, personalisation is simpler to realize since the personalisation information set can be located, fetched and used. However, the situation is more challenging when the user is moving and using different service instances installed in different servers from different devices. Most users are not aware of this usage of multiple service instances and will expect to use personalised services no matter where they are.

In this situation it is necessary to be able to locate, fetch the user's personalisation information set and use it in the adaptation of the service instance in use.

The third requirement is:

To enable personalisation, the user's personalisation information set shall be well defined and made available ubiquitously.

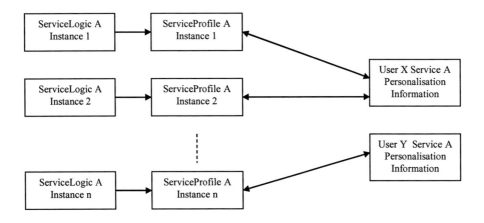

Figure 7 Separation of the Personalisation Information from the Service Logic

Consequently, a service should be decomposed into ServiceLogic and ServiceProfile and for each user there is a personalisation information set. Figure 7 shows *n* instances of service A and Personalisation Information for the service of two users X and Y. To personalise service A for user X, the ServiceProfile A will take the value from the User X Service A Personalisation Information.

It is then crucial that the service logic is able to use this personalisation information at initialisation to adapt itself to user's preference.

For that, the ServiceProfile must be equipped with an interface that allows the transfer of personalisation information, as stated in [7].

Such an interface must have at least two methods:
Set(personalisation_Info) – to assign the values of the personalisation parameters.
Get(personalisation_Info) – to extract the values of the personalisation parameters for saving and later usage.

Most current services and applications use local parameters for initialization and do not have the ability of querying and applying the user's personalisation information. To enable personalisation (see Figure 8), it is necessary to have an agent or software program, called **Personalisation Manager** that saves the original values of the parameters and uses the user's personalisation information to set these parameters. (One of the parameters must be the user's ID)

While using the service, the user may change both the presentation and functionality of the service. These changes must be captured and saved into the user's personal information. In addition, to preserve data consistency, the Personalisation Manager should also be equipped with a synchronisation function which keeps the Personalisation Information consistent and updated.

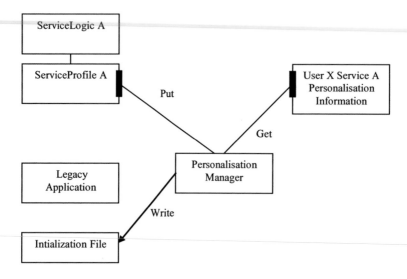

Figure 8 Personalisation with a Personalisation Manager

Figure 9 shows a UML class diagram illustrating the various components of the personalisation information.

Personalisation Information – This component represents the aggregate of all personalisation information. Ideally, this component should be provided as a shared component among as many service providers in as many service domains as possible.

User Personalisation Information – This component contains all personalisation information for one user.

Personal Information – This component contains personal information for a user. Examples of such information include addresses, phone numbers, credit card information etc. This information is in many cases unique, and more or less static, for each user.

Personal Generic Preferences – This component contains preferences that are generic to all services and service types/concepts. This could for example be font size settings and color selections.

Service Concept Personalisation Information – This component contains preferences that are generic to all services of a given type (henceforth the term service concept is used instead of service type). Thus, personalisation information that is common among several service implementations is moved out of the Personal Service Profile and into this component instead.

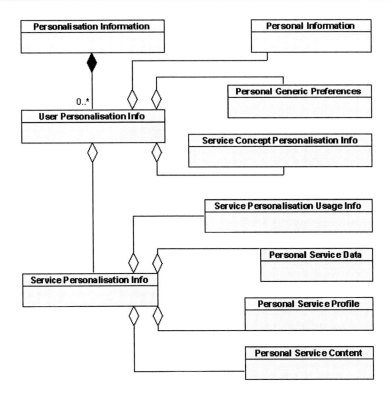

Figure 9 The components of the Personalisation Information

Service Personalisation Information – This component is the aggregate of all personalisation information that are specific to one service.

Service Personalisation Usage Information – This component contains information about the usage of a specific service.

Personal Service Data – This component contains service data that are related to a specific service.

Personal Service Content – This component contains service content that is related to a specific service.

Personal Service Profile – This component contains service profile that is related to a specific service.

To summarize, a mobile service can be decomposed in four components: Logic, Data, Content and Profile as shown in Figure 10.

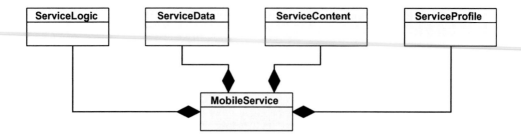

Figure 10 Composition of a mobile service

Service Personalisation and Service Continuity Requirements

Although Service Continuity and Service Personalisation are closely related, there are some differences in their requirements. Figure 11 captures these differences. As the figure illustrates, the primary goal of service continuity is to be able to interchange/substitute one service logic component for another. However, the primary goal of service personalisation is to be able to reuse the existing service content, service data and service profile. Reuse can mean either replicating/synchronizing the information or accessing it from the original location.

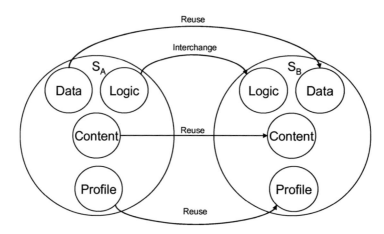

Figure 11 The major requirements of Service Continuity and Personalisation can be summarised by *interchange* and *reuse*

The combined requirements of service continuity and service personalisation are thus:

1. Service Continuity requires that it is possible to decide if one service logic can be substituted for another (and thereafter perform such substitution)
2. Both Service Continuity and Service Personalisation requires that it is possible to decide if service data of one service can be reused in another service (and thereafter transfer the data)
3. Service Personalisation requires that it is possible to decide if the service content of one service can be reused in another service (and thereafter transfer the content)

4.   Service Personalisation requires that it is possible to decide if the service profile of one service can be reused in another service (and thereafter transfer the profile)

To support requirement 1) requires an investigation of semantics and syntax of the two service logics involved. This is the primary focus of the next section. To support requirement 2)-4) also requires a study of the semantics and syntax of a service. The question is what the differences are of this semantics and the semantics used to support requirement 1).

To get an answer to requirement 1) the following question can be posed to the service logic in service S_B:

Question A: "What service do you provide?"

To get answers to requirements 2)-4) the following question can be posed, again to the service logic in service S_B:

Question B: "What (type of) data, content or profile are you demanding?"

A service must then in its semantic description keep information about the specific service it provides, as well as what data, content and profile it can make use of. These are not as unrelated as they first look. The data, content and profile will in most cases be dictated by the concept a service realises. If that is the case, based on the answer to Question A, it should be possible to deduce the answer to Question B. This will clarified in the later sections about Service Conceptualisation.

Equivalence between Services

One of the primary enablers of both Service Continuity and Service Personalisation is the ability to decide whether two services are equivalent or not. Service equivalence is determined by the combination of *syntactic equivalence* and *semantic equivalence*. That is, if two services are both syntactic and semantic equivalent, the services are said to be equivalent. In this section a formal approach is taken to illustrate these key points.

Services can be either atomic or composite. An atomic service is a service which can not be further divided into smaller components, whereas a composite service is a constellation of several atomic services. A composite service can be represented as a set (S_A) of atomic services ($s_1, s_2, \ldots$):

$$S_A = \{ s_1, s_2, s_3, \ldots, s_n \}$$

An atomic service consists of a pair of a semantic (C) and a syntactic (Z) property. C is considered to be a descriptive term and Z is considered to be a character string.

$$s_1 = (C_1, Z_1)$$
$$s_2 = (C_2, Z_2)$$
$$\ldots$$

The equivalence of two atomic services, s_1 and s_2 is then defined as:

$$s1 = s2 <=> \{C_1 = C_2 \wedge Z_1 = Z_2\}$$

Semantic equivalence (differentiated from equivalence by using two = symbols), which is weaker, is defined as:

$$s1 == s2 <=> \{C_1 = C_2 \wedge Z_1 \neq Z_2\}$$

Two atomic services are different if:

$$s1 \neq s2 <=> \{C_1 \neq C_2 \wedge Z_1 = Z_2\}$$

or if:

$$s1 \neq s2 <=> \{C_1 \neq C_2 \wedge Z_1 \neq Z_2\}$$

The two statements can be simplified into the following:

$$s1 \neq s2 <=> \{C_1 \neq C_2\}$$

Remember, however, that the overall service consists of several atomic services. Thus, the equivalence of the composite services S_A and S_B is defined by:

$$S_A = S_B <=> (\forall s \in S_A \; \exists t \in S_B \mid s = t) \wedge (\forall t \in S_B \; \exists s \in S_A \mid t = s)$$

In some cases, it might be sufficient that:

$$S_A \subseteq S_B <=> (\forall s \in S_A \; \exists t \in S_B \mid s = t)$$

This means that service S_B has equivalent atomic services for each of the atomic service in S_A, but it may include other atomic services which S_A does not have any equivalent atomic service of (i.e., S_A is a sub-set of S_B). For service continuity, this means that the transition from S_A to S_B is possible, but vice versa is not.

The equivalence of the syntactic properties Z_1 and Z_2 of s_1 and s_2 are simpler to determine than the equivalence of the semantic properties, because it is merely a question of character-wise comparison between the two strings Z_1 and Z_2. Finding a single difference means that the two are not syntactically equivalent.

To determine the semantic equivalence of elements ($C_1 = C_2$) requires more information. C_1 and C_2 must be well-defined and be elements in a set, call it Ω:

$$C_1 \in \Omega$$
$$C_2 \in \Omega$$

It is now clear that to be able to determine equivalence among two services it is necessary to define Ω. Ω is a set which contains all semantic terms for a given service domain. It is then such that:

$$X \subset \Omega$$
$$C_1 \in X$$
$$C_2 \in X$$

This means that the element A_1 is also an element of another set, call it X, which is a proper subset of Ω. This set X contains all terms that are semantically equivalent to C_1 (which could include C_2), and is an equivalence class.

Let us now consider two semantic equivalent atomic services *s1* and *s2:*

$$s_1 == s_2 <=> \{C_1 = C_2 \wedge Z_1 \neq Z_2\}$$

C_1 is the semantic of s_1 and s_2. Henceforth this semantic is referred to also as the *service concept* of s_1. Z_1 is the syntax of s_1 and Z_2 is the syntax of s_2. Henceforth this is referred to also as the *service implementation* of the two. Since Z_1 and Z_2 are different, it means that s_1 and s_2 instances of two different service implementations. To support service continuity and personalisation in this case it is thus necessary to define a transformation function $\mathfrak{I}$ which can be applied to Z_1 to obtain Z_2 (in general, $\mathfrak{I}$ can be applied to one syntax to obtain another):

$$s1 = s2 <=> \{C_1 = C_2 \wedge \mathfrak{I}(Z_1) = Z_2\}$$

The equivalence of two atomic services is thus defined by:

$$s1 = s2 <=> \{C_1 = C_2 \wedge (Z_1 = Z_2 \vee \mathfrak{I}(Z_1) = Z_2\}$$

Service Conceptualisation and Personalisation Relationship

Figure 12 displays a conceptualization of a service, and includes some terms that are used in the next sections.

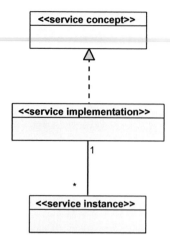

Figure 12 A conceptualization of a service

The following meaning and relationship among these terms are assumed:

Service Concept – An abstract idea of a service; e.g. the idea of a WWW browser or a word processor. This relates closely to C_n in the previous section.

Service Implementation – A realization of a service concept; e.g. Internet Explorer is a realization of the service concept WWW browser. A service implementation can realize one or more service concepts (for example, a WWW browser can typically open local text documents, and the WWW browser Opera is also an e-mail client). This relates closely to Z_n in the previous section.

Service Instance – The unique instance of a service implementation in a specific location; e.g. the installation of Internet Explorer on a PC. This releates closely to s_n in the previous section.

The primary enablers of personalisation are the elements that contain information linking *users* to *services*. These are elements of the Personalisation Information Space as previously defined in [7]. In Figure 13, there is one personalisation relationship for each service instance. Thus, this is defined as the concept of *Local Service Personalisation* (to indicate that the personalisation is *local* to the specific *service instance*):

Local Service Personalisation – Personalisation is constrained/limited to cope with each individual service instance, and thus also with each service implementation.

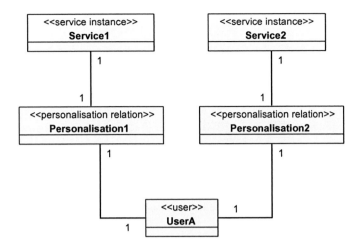

Figure 13 Local Service Personalisation

In Figure 14, there is one personalisation relationship shared by a number of service instances. This is the concept of Global Service Personalisation.

Global Service Personalisation – In Global Service Personalisation, several service instances can be personalised by the same personalisation information.

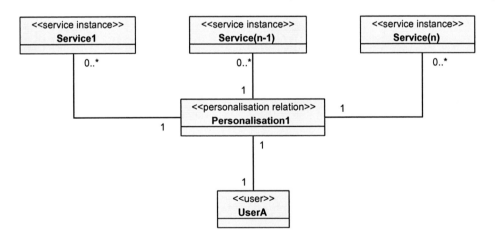

Figure 14 Global Service Personalisation

However, the two previous models illustrating the personalisation relationship do not consider the service implementation, or rather *different* service implementations. Figure 15(a) shows the concept of different service instances of the same implementation being personalised, while Figure 15(b) shows the case where different service instances of different service implementations are personalised through the same personalisation relationship.

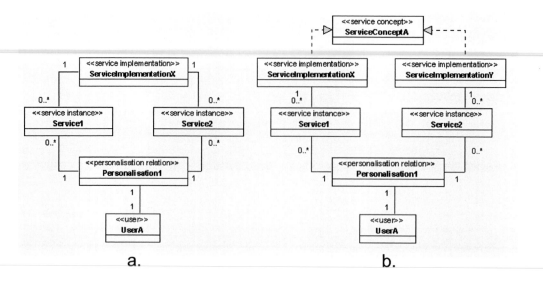

Figure 15 Global Service Personalisation (a.) and Global Cross-Service Personalisation (b)

Figure 15(b) shows the ultimate goal, where personalisation is independent of both service instance and service implementations. Note also that both implementations in Figure 15(b) are realizations of the same concept; for the most cases, this is a requirement, because it seldom makes sense to use personalisation information from a service of one concept, in another service realising another concept (e.g. using an MS Word document with an mp3-player would in most cases yield noise). However, generic preferences are possible, so it is also necessary to consider a Global Cross-Service Concept Personalisation, as depicted in Figure 16.

The condition for both Global Cross-Service Personalisation and Global Cross-Service Concept Personalisation is the ability to compare and conclude about the equivalence or difference between the service implementations as dictated by the previously defined *Service Equivalence*.

Organisation of Personalisation Information

To enable the comparison between service implementation and personalisation information, it is necessary to define an ontology specifying the organizational structure between services. It is necessary to define relationship between services and the rules to navigate in the structure. For example, given two services it should be always possible to determine the relation between them (equivalent, different, subset, etc.)

Figure 17 shows how different personalisation information elements (represented as stereotyped <<PIE>> in the UML class diagrams) are associated with a specific service implementation through a service concept. It should be possible for a service implementation to both inherit a specification from the service concept, and to provide an additional specification (i.e., a direct association from the <<*service implementation*>> to the specific <<PIE>> instance).

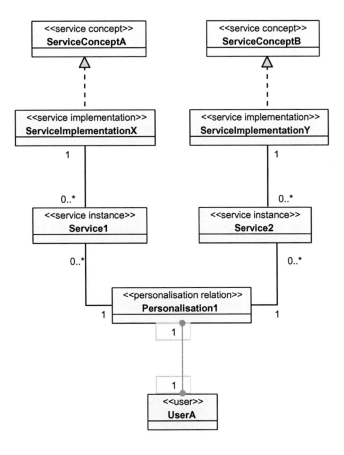

Figure 16 Global Cross-Service Type Personalisation

Applying Service Equivalence

This section performs a study of how the previously covered topics can be used in practice. Without going into too much detail here, the semantic properties (e.g. C_1 from previous section) may include references to (as defined earlier in this chapter):

- Service data
- Service content
- Personalisation information

According to the previous section, combined with the definition of the mobile service composition, it is now possible to represent a service as S_A and S_B in Figure 18. One of the ideas applied here is that service logic, service data, service content and service profile all can be exposed as atomic services themselves, as proposed in [11].

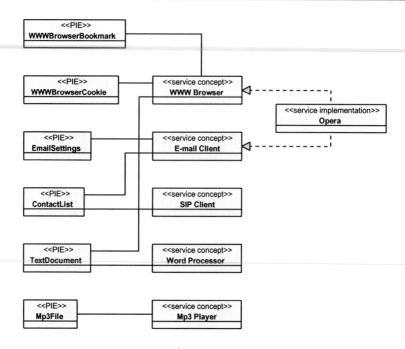

Figure 17 Personalisation Information Elements (PIE) associated to service implementations through service concepts

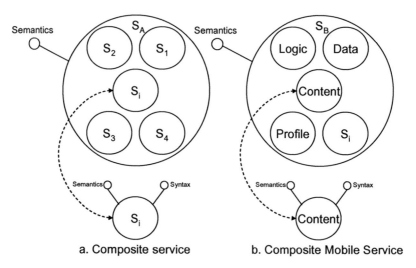

Figure 18 Composite services expose semantics, whereas atomic services expose both semantics and syntax (a). In mobile services, the content is exposed as an atomic service (b).

The process of determining service equivalence and achieving service continuity is outline in Figure 19.

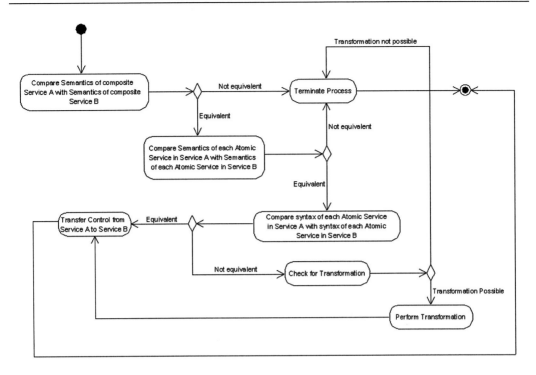

Figure 19 Outline of the process for determining service equivalence and achieving service continuity and personalisation

The next sub-section discusses how personalisation can benefit from employing ontologies and in detail how personalisation information can be mapped to ontologies. It is important to realise that the "Content" in Figure 18 also can represent personalisation information.

OWL Ontology Specification

The model of personalisation information described in Figure 17 in a previous section is only a visual representation of the personalisation information elements and their relationships to service concepts and implementations. To be machine processable, the visual model must be transformed into a textual representation.

The OWL Web Ontology Language [12] is based on RDF [13], and is a language for defining and instantiating Web ontologies. Below is the specification of the personalisation information ontology in abstract OWL syntax [14][15]. Only one service concept is covered (the WWW browser) in this example ontology.

```
Ontology(

Annotation(rdfs:comment  "Ontology  used  for  personalisation  of  mobile
services")

Annotation(rdfs:label "Personalisation Ontology")
Annotation(owl:imports http://www.ongx.org/service₁)
Annotation(owl:imports http://www.ongx.org/service₂)
...
...
```

```
...
Annotation(owl:imports http://www.ongx.org/service_n)

ObjectProperty(personalised-by inverseof(personalises))

ObjectProperty(personalises domain(serviceConcept))

ObjectProperty(realised-by inverseof(realises))

ObjectProperty(realises domain(serviceConcept))

Class(serviceImplementation partial annotation(rdfs:comment "A specific service
implementation"))

Class(serviceConcept partial annotation(rdfs:comment "The abstract concept of a
service"))

Class(piElement partial annotation(rdfs:comment "A personalisation information
element"))

Class(piSet complete annotation(rdfs:comment "The set of all personalisation
information elements") piElement)

 Class(WWWBrowserBookmark partial piElement)
 Class(WWWBrowserCookie partial piElement)
 Class(TextDocument partial piElement)
 Class(Mp3File partial piElement)

 Class(WWWBrowser partial serviceConcept
 restriction(personalised-by WWWBrowserBookmark WWWBrowserCookie TextDocument))

 Class(Opera partial serviceImplementation
 restriction(realises WWWBrowser))
 )
```

The ontology starts by including already existing ontologies (using *owl:imports*). The ontology should as such be expandable, and could in theory include ontologies developed by various service providers. With reference to the section on equivalence between service, the set Ω has now been defined as:

$$\Omega = \{ \; WWWBrowser, \quad Opera, \quad WWWBrowserBookmark, \quad WWWBrowserCookie, \\ TextDocument, \; Mp3File, \; ... \; \}$$

An example of the set X mentioned in the same section could be:

$$X = \{ \; WWWBrowser, \; Opera \; \}$$

This signifies that Opera is a WWWBrowser, and that Opera can be personalised by the same personalisation information elements as a WWWBrowser. However, the OWL ontology above defines more than the equivalence classes. It also defines any other relationship between entities.

In a real implementation, OWL-Lite, OWL-DL or OWL-Full would be used to build the ontologies. However, due to its undecidability, OWL-Full might not be appropriate. There $\exists$ tools for parsing and reasoning about OWL-Lite and OWL-DL, e.g. the Pellet OWL-DL reasoner [16]. The use of ontologies can make sure that some of the most important requirements of next generation mobile services are supported, i.e. *Service Personalisation* and *Service Continuity*.

Conclusion

This chapter has covered topics related to future mobile services: from the high-level requirements of future mobile services, through the more formal treatment of service equivalence and finally to the support of the identified requirements and challenges by the use of ontologies based on standard languages like OWL. First, the current architectures of mobile services were described and their characteristics, strengths and limitations were illustrated. The main components of a generic mobile service were identified and elaborated. Requirements relating to Service Continuity and Service Personalisation were discussed next. Service personalisation was discussed in detail, and an information model was elaborated. The chapter then took a required detour to consider service equivalence in a more formal manner, before proceeding with the more practical issues of realising service equivalence logic and defining ontologies that provide support for service continuity and service personalisation.

References

[1] Blom, J. (2000), "Personalisation – A Taxonomy", *Conference on Human Factors in Computing Systems* (CHI), Hague, Netherlands, April 1-6, 2000, ISBN: 1-58113-248-4

[2] M. Mouly and M.-B. Pautet. *The GSM System for Mobile Communications*. M. Mouly and M.-B. Pautet, Palaiseau, France, 1992.

[3] 3rd Generation Partnership Project 3GPP: Technical Specification TS 03.40 V7.5.0 (2001-12) Technical Specification Group Terminals; *Technical realization of the Short Message Service* (SMS) (Release 1998)

[4] Jørstad, I., Dustdar, S., van Do, T. (2004). "Towards Service Continuity for Generic Mobile Services". *The 2004 IFIP International Conference on Intelligence in Communication Systems* (INTELLCOMM 04), Bangkok, Thailand, 23-26 November 2004. ISBN: 3-540-23893-X

[5] Jørstad, I., Dustdar, S., van Do, T. (2005). "Service-Oriented Architectures and Mobile Services", *Ubiquitous Mobile Information and Collaboration Systems* (UMICS2005), Porto, Portugal, 13-14 June 2005,

[6] Jørstad, I., Dustdar, S., van Do, T. (2004). "Service Continuity and Personalisation in Future Mobile Services". 10th International IFIP *Workshop on Advances in Fixed and Mobile Networks* (EUNICE 2004), Tampere, Finland, 14-16 June 2004. ISBN: 952-15-1187-7

[7] Jørstad, I. & van Do, T. (2004). "Personalisation of Future Mobile Services". *9th Interational Conference on Intelligence in service delivery Networks* (ICIN), Bordeaux, France, October 18-21, 2004

[8] ETSI/3GPP. (2005). "Universal Mobile Telecommunications System (UMTS); *Service requirements for 3GPP Generic User Profile* (GUP);Stage 1 (3GPP TS 22.240 version 6.5.0 Release 6)". ETSI, January 2005

[9] ETSI/3GPP. (2005). *"Universal Mobile Telecommunications System* (UMTS);3GPP Generic User Profile (GUP) requirements;Architecture (Stage 2) (3GPP TS 23.240 version 6.7.0 Release 6)". *ETSI*, March 2005.

[10] ETSI. (2005). "Human Factors (HF); User Profile Management". *ETSI*, August 2005

[11] Jørstad, I., van Do, T. (2005). "A Service-Oriented Architecture Framework for Mobile Services", *Advanced Industrial Conference on Telecommunications* (AICT2005), Lisbon, Portugal, 18-22 July 2005, ISBN: 0-7995-2388-9

[12] Smith, M.K., Welty, C. & McGuiness, D.L. (2004). *"OWL Web Ontology Language"*, W3C Recommendation, February 10, 2004, http://www.w3.org/TR/owl-guide/

[13] Beckett, D. (ed.) (2004). *"RDF/XML Syntax Specification (Revised)"*. W3C Recommendation, 10 February 2004, online: http://www.w3.org/TR/rdf-syntax-grammar/

[14] Antoniou, G. & Harmelen, van F. (2004). *A Semantic Web Primer*. MIT Press, 2004, ISBN: 0-262-01210-3

[15] Patel-Schneider, P.F., Hayes, P. & Horrocks, I. (2004). "OWL Web Ontology Language Semantics and Abstract Syntax", *W3C Recommendation*, February 10, 2004, http://www.w3.org/TR/2004/REC-owl-semantics-20040210/

[16] Parsia, B. & Sirin, E. (2004). "Pellet: An OWL DL Reasoner", *3rd International Semantic Web Conference* (ISWC2004), Hiroshima, Japan, November 7-11, 2004, Springer-Verlag, Lecture Notes in Computer Science, ISBN: 3-540-23798-4

In: Mobile Multimedia: Communication Engineering ...   ISBN 1-60021-207-7
Editor: I.K. Ibrahim and D. Taniar pp. 43-62   © 2006 Nova Science Publishers, Inc.

Chapter 3

PROTOCOLS AND ARCHITECTURES FOR MOBILE COMMUNICATIONS IN MODERN IP NETWORKS

Giuseppe De Marco* and Leonard Barolli[†]
Department of Communication and Information Engineering
Fukuoka Institute of Technology, Japan

Abstract

The problem of mobility of IP hosts dates back to early 90's, when Mobile IP (MIP) was proposed as solution for Mobile Nodes (MNs) which use the same IP address wherever they move into. MIP was conceived for macro-mobility, i.e. the movement among different administrative domains. But now it is also used for movements among different technologies. The emergence of wireless networks in the last decade has prompted the research towards the problem of micro-mobility, i.e. the frequent change of Point Of Attachment (POA) to the network. In fact, in wireless networks the physical dimension of the subnetwork affects the rate of hand-off, which in turn bears on the Quality of Service (QoS) of common mobile communications, such as Web browsing and Voice Over IP. The wireless IP networks differ from classical cellular networks, where a managed network infrastructure is present and where network protocols are standardised or engineered in ad hoc fashion. Nowadays, we can count several proposals about the micro-mobility problem in IP networks. Although we cannot say when the expectations of these technologies will be realised, we think that some fundamental properties of IP networks should be met in order to make the micro-mobility solutions feasible. One of the design guidelines of IP networks has been the end-to-end (e2e) paradigm, which means moving complexity into the terminals instead of adding more components inside the network. Another basic aspect of the IP networks is theirs packet based nature. For this reason, the latencies of subnetwork hand-off are critical factors, because the IP address acts as localisation of the MN and its identification as well. This means that the notification of the IP addresses is necessary whenever a movement into a different network takes place. In this chapter, we try to summarise current proposals by comparing the performance and architectural problems of every solution. We begin this overview by introducing a

---

*E-mail address: demarco@fit.af.jp
[†]E-mail address: barolli@fit.af.jp

general taxonomy which permits to better identify pros and cons of current IP micro-mobility protocols. This classification should help understanding that e2e solutions do not cope with simultaneous movements of mobile nodes, whose probability is not negligible, especially in the case of high-speed hosts. Hence, a mixed scheme could be better. Along with this classification, we present some possible improvements of current proposals. In particular, we design a new pseudo e2e scheme, namely the smartAR scheme, which reduces network complexity by exploiting existing components, for instance the Access Router (AR) which is always present within an IP network. Furthermore, this scheme endures both single and simultaneous movements scenario. We enrich this overview on IP micro-mobility protocols with a simple analysis of the hand-off latencies.

Keywords: IP micro-mobility, Performance Evaluation, Hand-offs, Wireless IP Networks, SIP, SCTP.

1 Introduction

A Mobile Node (MN) is connected to the IP network through a Point of Attachment (POA). The movement of an MN across different subnetworks is called micro-mobility. This is the complement of the macro-mobility case, where the MN selects a new POA which belongs to a different network and/or a different technologies. Sometimes the terms horizontal and vertical hand-off are used for the micro-mobility and macro-mobility hand-off, respectively. The micro-mobility is more delicate in wireless networks, because the change of subnetwork can be more frequent due to the reduced size of the radio cell. Mobile IP was the first attempt to solve the problem of macro-mobility and micro-mobility as well. Mobile IP needs special localisation agents, e.g. Home Agent (HA) and Foreign Agent (FA), in order to correctly forward data traffic towards the current location of the MN. In Mobile IP, the MN is assigned a permanent and globally routable home address. Many other protocols have been proposed in the last years to handle the micro-mobility. For example, we can cite Session Initiation Protocol (SIP) which is the model of every application based micro-mobility protocol. Also in SIP, we need localisation servers to retrieve the exact position of the MN in terms of its reachable IP address. Other solutions are based on transport layer optimisation. For example, TCP-M [20] and m-SCTP [9, 16]. Recently, end-to-end solutions enriched with Peer-to-Peer (P2P) communication for distributed naming system have been proposed [5]. However, every solution has its advantages and its disadvantages. The right choice depends much on the type of application the protocols are designed for. Easiness of architecture deployment can also be a success factor, indeed. For this reason, end-to-end solutions are always preferred. On the other hand, full e2e management of IP addresses can be harmful when we consider the simultaneous movements case.

In this paper, we give a general methodology to classify these proposals. The aim of this methodology is twofold. Firstly, it serve as a tool to inspect possible architectural and/or protocol redundancies. Secondly, it suggests how to design new solutions in a comprehensive and clear way. In fact, by this methodology we design smartAR, a solution which is quasi-e2e, because the difficulties of mobility management are moved from the MN to the Access Router (AR), that is the last hop before the MNs. In our opinion, this solution is feasible, because with a little updating of the AR, which is always present, we obtain a

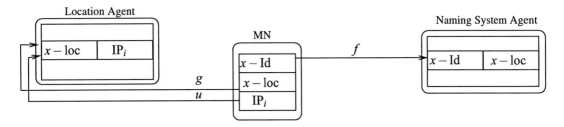

Figure 1: Conceptual view of mobility functions.

scheme which handle both single and simultaneous movements. Furthermore, the hand-off latencies are reduced. The localisation system is left to the application.

The rest of the chapter is organised as follows. In Section 2, we describe the functions of our taxonomy which is used to classify current micro-mobility protocols in Section 3. In Section 4, we summarise fast-hondover protocols in IEEE 802.11.x networks. In Section 5, we describe the design guidelines of smartAR and in Section 6 a simple performance analysis of hand-off latencies is given. Conclusions and discussion of presented work are in Section 8.

2 Mobility Functions

In an IP network, every host can be thought as a triple of values. The first value, x-Id, is an alphanumeric string which represent the identity of the host. The second value represents the location of the host, x-loc. For example, x-loc can be the address of the server storing the exact location (e.g. IP address) of the (mobile) host. The last value is the standard IP address. Accordingly, the IP micro-mobility requirements are easily summarised by the following functions, explained also in Fig. 1

- The identification function, $f : x\text{-Id} \to x\text{-loc}$, which is used for example when an originating entity, e.g. a Correspondent Node (CN) or another MN, wants to communicate with the MN (data are transmitted to or received from it). The originating entity knows the identity of the destination, x-Id (e.g. in SIP, x-Id=¡giuseppe@sip.fit.ac.jp¿), and needs its locator, x-loc. For instance, f function is needed whenever a naming service is being used, as in the case of Domain Name System (DNS).

- The localisation and forwarding function, $g : x\text{-loc} \leftrightarrow \text{IP}$, which maps the x's locator to a reachable IP address and vice versa. The g-function is invoked in order to localise the destination IP address, which can vary over time. This function can be executed several times, because the exact address of the MN can be retrieved by querying many servers. For example, if we have two location servers, X_1 and X_2, in two different domains, the CN first executes $f(\cdot) = X_1$, then $g_{X_1}(x\text{-loc}) = X_2$ is executed in the X_1's location server, and finally $g_{X_2}(x\text{-loc}) = IP_{x\text{-loc}}$ is executed inside the X_2's server. Thus, we have executed two times g function, but in different "places". To avoid confusion, we should use a subscript for g function to discriminate the place where g is being executed. As in the previous example, g_{d_1} means that g is executed in the domain d_1. We will omit this subscript whenever it is not strictly required. In general, $g(\cdot)$ is not injective, that is g can have multiple IP addresses associated to its locator. This is the case of multiple network interfaces or multiple IP addresses assigned to the same Network Interface Card (NIC). On the other hand, this is also the case of branching searching process of the user for particular application, e.g. the MN is registered with two location servers.

- The updating function $u : (x\text{-loc}, \text{IP}_{\text{old}}) \rightarrow (x\text{-loc}, \text{IP}_{\text{new}})$. Although this function is indeed a simple storing operation, we maintain the term. This function means that the location information stored somewhere in the network must be updated. In this case, $x - \text{loc}$ acts as a pointer to the storing place of that information. The u function is usually executed by MN when its IP address has to be changed. This function directly affects the cost of a handover, which right happens when the MN's IP must be changed. To be precise, the cost of u is composed of two parts: detection cost of a movement and updating cost of IP address. Detection movement concerns the waiting of a proper event which makes MN aware of the movement. This delay usually depends on the technology used. In general, without special L2-type trigger signals, this delay depends on the signalling messages or subnetwork advertisements (see Section 3.1). It is clear that to be feasible every performance gain of IP micro-mobility protocols should take into account also the movement detection time. The detection time in turn strongly depends on the technology used. For example, in cellular networks this operation is entirely taken at the L2 layer, with carefully engineered radio techniques (for a review see [15]). We will see in the subsequent sections that all IP micro-mobility proposals try to minimise the cost associated with this function. In a centralised solution, this function will also update the MN's IP_{old} address at the locator agent, e.g. the SIP server.

These functions are sequence of operations. How these functions are performed depends on the protocol. Our first contribute is to use these functions in view of a comprehensive classification of micro-mobility protocols.

3   A Menagerie of Protocols

In this section, we describe most popular IP micro-mobility protocols by means of the taxonomy previously discussed. We chose just few examples, because the characteristics of other protocols are similar. We use the term mobile host and mobile node interchangeably. In all schemes, we assume that the handover movement detection phase is composed by L2 and L3 signalling procedures. All these procedures are embodied in what we call Black Out Phase (BO). We use the terms L2 and L3 to mean layer 2 (link layer and radio layer) and layer 3 (network layer), respectively. In what follows, single movement means that one and only one of two hosts of a communication performs a movement. The term simultaneous movement points out the event in which both MNs moves, and their combined movement is critical in the sense that g function uses IP addresses not yet updated. In other words, a movement of an MN takes place when the IP address of correspondent MN is being updated. In this section, we also discuss other notable protocols, such as Cellular IP [23],

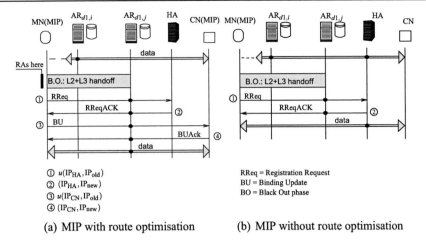

(a) MIP with route optimisation   (b) MIP without route optimisation

Figure 2: MIP operations in the case of in-session movement

TeleMIP [2], Dynamical HMIP [11], TCP Migrate (TCP-M) [20] (see Section 3.4, for a brief description).

3.1   MIP and HMIP

In Mobile IP, every node has a global routable address, called home address. This address is stored in a particular routing agent called HA. Thus, the HA stores $x - \text{loc}$. If the MN owns also an alphanumeric identity, $f(x - \text{Id}) = (x - \text{loc})$ is performed by querying a DNS. Otherwise, we have $x - \text{Id} = x - \text{loc}$. Whenever a movement into a new subnetwork is detected, the MN starts a sequence of operations in order to get a new address called Care-of-Address (CoA). The MN executes u function. This CoA represents the IP address of a special routing agent in the visited subnetwork called FA. The CN sends data traffic to HA, which in turn sends packets towards the FA by using IP-in-IP encapsulation. The FA decapsulates the packets towards the MN. In MIPv6 this mechanism is simplified because the MN has a globally routable address [13]. Although FA is not strictly necessary, some kind of AR must be present. In fact, movement detection and address configuration, i.e. the selection of a new CoA, would not be possible without the AR. In Fig. 2, without confusion and for simplicity we used the general term AR only. L3-type movement detection is part of BO phase and needs special signalling messages called Router Advertisement (RA). These messages are sent periodically over the wireless link to advertise the subnetwork identity, i.e. its link prefix. In MIPv6, upon detection of a new subnetwork, the MN can automatically formulates a new CoA, usually called *tentative* or *proposed* CoA [10]. Since there is a probability of address conflict, a procedure called Detection of Address Duplication (DAD) is performed. However, it is recognised that this probability is very small. In MIPv6, the Return Routability Test (RRT) is another important feature which decrease the transmission delay between two peers and avoid triangle routing. The new CoA is then registered back to the HA. Let us note that u function is executed at every L3-type movement. For high-speed users this function can be very costly. To avoid frequent and costly CoA registrations, several enhancements of MIP have proposed a hierarchical architecture. Hierarchical MIP (HMIP) has been the first solution aimed at transforming MIP in a true micro-mobility pro-

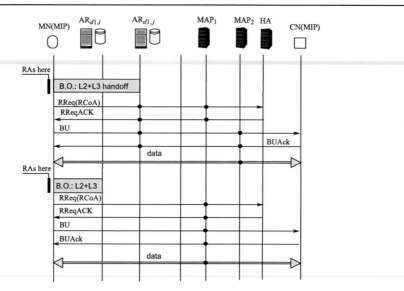

Figure 3: HMIP in case of movement into a new region.

tocol [21]. The logic is simple: one protocol for handling local movements and one for macro-movements. For instance, the network domain is divided in regions and hierarchies. Every region has a Mobility Anchor Point (MAP) which is responsible for a specific group of subnetworks. Multiple hierarchies are possible by assigning many MAPs to another MAP, and so on. The MNs register their CoAs, called Local CoA (LCoa) into the MAP and perform registrations back to the HA if and only if a change of region is detected. At the first POA to the network, MN registers MAP's address as CoA into the HA. This CoA is called Regional CoA (RCoA). RCoA are set up whenever RAs received by MN contain a MAP option. By this way, MN can detect a movement into another region. Packets towards MN are intercepted by HA, which in turn forwards encapsulated packets to the MAP the MN has been serving with. The MAP forwards (by decapsulating or not) packets towards the right FA [1]. By this way, the main part of L3 handovers latency is due to the registration into MAPs. Other extensions of HMIP consider the addition of paging mechanisms. However, we must add new components in the network. Moreover, neither MIP nor HMIP can handle simultaneous movement. For details about MIP in simultaneous movements see [24].

3.2   End to End Protocols

3.2.1   SCTP: Pure e2e Solution

MIP and its variants are not e2e protocols. They require a large scale deployment of HAs and/or MAPs. Alternatively, a very simple e2e approach is possible at the transport layer. Whenever an MN acquires a new CoA, it executes the u function back to the other host. Home registrations are not strictly needed and the handover process is faster. This benefit is particularly attractive for MNs close to each other (for instance, in the same network domain). An example of this approach is the mobile Stream Control Transmission Protocol

---

[1]In MIPv6 the FA is not strictly necessary.

(m-SCTP) [9, 16, 22]. Besides other features like multistreaming and multihoming, SCTP can offers a dynamic address management of active connections or rather association in SCTP terminology. In fact, by means of ASCONF messages an MN can notify the other host of its new IP address. In particular, a new available IP address for the current association is inserted in an ASCONF message, called ADD-IP. Similarly, an IP address can be deleted from current association. In Fig. 4, a typical example of mobile SCTP is shown. The association is created by a 4-way handshake; after handover is accomplished (i.e. after correctly configuring a new address), MN executes u by sending an ASCONF-ADDIP to the correspondent host. In this case, u function uses CN's IP as $x - loc$. Besides its true and attractive end-to-end peculiarity, another important aspect of this solution is that we are not asked to re-open a connection socket. In fact, if we used TCP, we would have had to tear down the current socket and reestablish the connection. However, two drawbacks are present in this solution. Firstly, ASCONF has been designed for path backup purposes in wired networks. In wireless networks, some performance degradation can arise. In [3], the authors show some performance improvement of loss recovery phase of SCTP during handover. On the other hand, as in MIP, also here simultaneous movement cannot be correctly managed. In fact, in simultaneous movement the destination address of an ASCONF-ADDIP message and the current IP address of the other host are different. See for example Fig. 5. However, end-to-end solutions are very fast and inexpensive. Let us note that m-SCTP cannot perform f function, that is we cannot locate the MN without support of some naming service and/or host lookup system. This was expected because, in general, transport protocol are unaware of the type of application. The f function is a requirement of the application. We will address this aspect in Section5.

3.2.2 SIP

SIP has been studied for several years as an application protocol for signalling procedures in multimedia oriented communications [8, 17]. A number of other application protocols arose in the last years, especially with the success of working softwares like Skype© [19]. SIP is basically a transactional protocol, within which two kind of transactions can take place: INVITE transaction and non-INVITE transactions. The former is executed whenever an host wants to establish a communication towards another host. The latter group comprises all other transaction, like registration transaction and communication tear down. Transactions in SIP are coded by alphanumeric strings which forms the SIP messages. In particular, INVITE is also the name of a particular field of the message sent during an INVITE transaction. Similarly, other transactions are coded into different text messages, as in Fig. 4-b. When a handover happens, SIP uses the Re-INVITE message in order to inform the correspondent host of new IP address. A REGISTER message is sent after the Re-INVITE. If the Re-INVITE is sent end-to-end, that is without passing through a SIP server, we fall into the SCTP problem: simultaneous movements are not properly handled. On the other hand, if we force SIP messages to traverse a SIP server, we can handle simultaneous movements. To be more clear, consider Fig. 5 and SIP signalling. We have two MNs, MN_1 and MN_2, respectively, belonging to two different domains. $SIP-P_1$ and $SIP-P_2$ are the proxy servers of the two domains. Reliability of signalling messages is based on timeouts[2]. After exe-

---

[2]In the case of reliable transport protocol as TCP similar explanation holds.

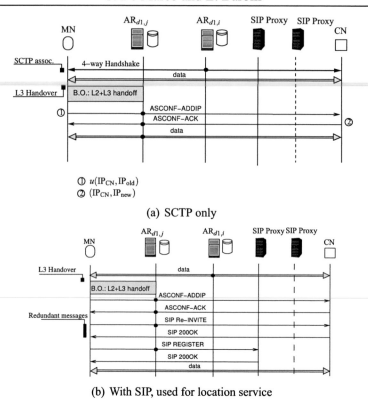

(a) SCTP only

(b) With SIP, used for location service

Figure 4: SCTP used for dynamical address management during handovers.

cution of u function at point 1, if SIP server does not receive any response, it will re-send the message after the expiration of a back-off timer. This can be too long, even for a single retransmission ($\sim 500ms$). In order to speed up function u, SIP-$P_{1,2}$ should immediately send pending messages whenever they receive a REGISTER message. This means that SIP servers must be state-full. Moreover, we should guarantee that all SIP messages are routed through SIP Severs.

3.3 EMIPv6

This protocol and is based on a mixed architecture. In fact, the searching functions like g is executed by means of a P2P network, that is the function g is repeatedly executed inside nodes of a distributed network. This architecture is surely scalable and robust due to the use of the P2P paradigm. The details of this solution are shown in [5]. The authors solve the problem of simultaneous movements by means of a distributed Subscription/Notification (S/N) architecture, which resembles the SIP based location service. The only difference between the SIP architecture and EMIPv6 is that this S/N system is P2P based. Since the g and u functions are executed inside peer nodes, there is no danger of signalling messages loss, which cause the anomalies of simultaneous movements. The success of the address resolution process is guaranteed by the fact that the neighbours of the destination host always know its right or rather most updated address. The authors have showed that with 80000 nodes the median of the resolution latency is around $400ms$. This solution is very similar to

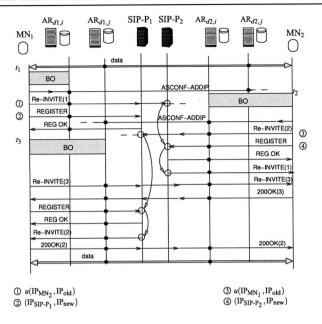

Figure 5: SCTP and SIP in the case of simultaneous movement.

existing and popular softwares, like Skype© [19], and it is a proof that P2P architectures are viable and scalable solutions to difficult problems of modern telecommunication networks.

3.4   Other Protocols

There are other proposals which are close to hierarchical scheme of HMIP. In Cellular IP [23], the network is divided in many interconnected cellular nodes, and every node maintains a routing cache for every MN. Macro-mobility or inter-domain mobility is fulfilled by MIP. A gateway connects the cellular domain to the Internet. Hop-by-hop packets are exchanged among cellular nodes for updating the routing caches. In Cellular IP, one can use also semi-soft hand-off by means of bicasting. Bicasting is a technique that permits to forward data traffic to old POA as well as to the new one. In this way, packet losses are reduced, but packet duplication and re-ordering can not be neglected. A discussion of this effect can be found in [1], where the authors show the effects of circular buffer length on the packet re-ordering and hand-off delays. Another feature of Cellular IP is the paging support. TeleMIP has two-level hierarchy; gateways can be connected to several FA. By this way, gateways are chosen by means of load balancing policies [2]. TCP Migrate (TCP-M) [20] was conceived in order to maintain TCP connection during handover. TCP-M is based on DNS updates to execute mobility functions. Its structure resembles that of SCTP, and then it is considered an e2e solution. Dynamical HMIP is a variation of HMIP [11]. In this protocol, the MN registers new CoAs back to HA after traversing a number of FAs, which represents a threshold after which a home registration is triggered. The authors of [11] focus on the exact relation between this threshold and the overall hand-off cost. However, although the benefits of this technique depend on the architecture of the network, we note that, in mobility, an MN traversing more than two FAs during a connection is quite rare.

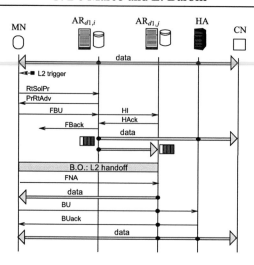

Figure 6: Procedures of fast handover protocol.

4 Fast Handover Protocols

As discussed above, in all IP micro-mobility schemes, the movement detection time is a critical factor of the overall handover cost. Making handover before it happens can speed up the configuration of address and other parameters of the MN. Fast Handover Protocols (FHP) allow to execute u function while the MN is still in the old subnetwork. To this aim, MN should be supported with some L2 functionalities because MN should be able to hear signals from two different POAs. The set of signalling messages and operations of all FHP proposals resembles to that of Fast MIPv6 (FMIPv6) document [10]. In Fig. 6, an MN is made aware of a new neighbour subnetwork by means of special L2 signals (see next section). Afterwards, the MN sends a Router Solicitation for Proxy advertisement (RtSolPr) message to its current AR, that is $AR_{d1,i}$. The Proxy Router Advertisement (PrRtAdv) message sent by the AR contains information about the new discovered subnetwork, in particular it contains its address prefix. Accordingly, MN can formulate a tentative CoA which is confirmed by Fast Binding Update (FBU) and FBack messages. The ARs should be able to communicate to each other.

4.1 Fast Handover over 802.11 Networks

The feasibility of fast handover protocols depends on L2 triggers. In current standard technologies, like IEEE802.11, listening to two different APs with same NIC is not allowed. The problem in IEEE802.11 is that the connection to an AP entails three phases: channel scanning, channel switching and authentication. Since IP packets cannot be transmitted and/or received while the link layer is in the scanning phase, the MN cannot send any signalling message to the new AR without losing the current connection. This means that also the so called *make-before-break* hand-off cannot be performed with this technology. However, the interest in fast hand-off in IEEE 802.11.1x is witnessed by the activities of standardisation institutions like the IETF [12], which are considering results from current proposals.

To overcome the constraint of the link layer, a technique to interleave scanning phase of neighbour APs (and then AR) and data transmission has been proposed. This technique uses NICs of the MN in a multiplexed fashion; during the scanning phase, MNs places the current APs in power-saving mode [14] and try to scan the radio medium to find neighboring APs. The switching period is computed in order to minimise the reception latency of a beacon message from neighbour APs. The experimental results of this technique showed that during the hand-off between two APs, the packet loss is negligible as well as the delay jitter of received packets..

Another form of L2 trigger can be realised by means of a different approach. In [6], in order to decrease the hand-off rate between WLAN and WWAN, the authors suggest to use a smooth version of the samples of the received signal strength. In fact, when the MN is going out of the current serving WLAN cell, the sequence of these samples is always decreasing. By using the fundamental term of the Fast Fourier Transform (FFT) of this sequence, $X(1)$, it can be shown that:

$$E[X(1)] < 0, \tag{1}$$

where

$$X(1) = \sum_{n=0}^{N-1} x(n) \sin\left(-\frac{2\pi n}{N}\right), \tag{2}$$

here N being the length of the samples sequence. The authors compare this value with a pre-computed threshold in order to decide whether a hand-offs procedure should be executed. By this way, the authors are able to reduce the number of unnecessary hand-offs and predict the true ones as well. However, if the firmware of the NIC is not open, some cross-layer techniques should be used because the computation of the FFT should be done at higher layers. In [6], the solution to the problem of simultaneous movements is similar to [5]. Another idea which has been considering for future versions of IEEE.802.11x is that of neighbour graphs. In [18], the authors suggest to build a neighbour graph which *stores* the hand-off relations among adjacent APs. By means of this graph, the MNs can reduce the L2 channel scanning phase, because they know in advance the right channel to use.

5   A new Proposal: smartAR

Before introducing our proposal, let us summarise above mentioned protocols with help of Table 1. A ($\cdot$) means that function could be executed by the protocol, though not conceived by its base version. As first observation, almost all protocols provide same functionalities. In particular, if we have a MIP architecture and we use SIP, then redundant operations will happen. In fact, in this case both SIP and MIP performs the same function[3]. On the other hand, full and e2e solutions are possible, as EMIPv6, but the main obstacle is the cost of g which depends on the searching time of P2P mechanisms. Moreover, in EMIPv6 the searching phase is based on arbitrary mechanisms. In our opinion, the optimal solution is reached when there are no redundancy among protocols and network complexity is minimal. From this point of view, one can assert that SIP+mSCTP could be a candidate as optimal solution.

---

[3] In MIP g is executed also for packets forwarding, while in SIP it is executed for localisation purposes only.

Table 1: Comparison of IP micro-mobility protocols.

Protocol	L2 trigger	e2e	f	g	u	Mobility Agents	Sim.Movements
MIP	no	no	(•)	•	•	1+	no
HMIP	no	no	(•)	•	•	2+	no
DHMIP	yes	no	•	•	•	2+	no
FMIPv6	yes	no	(•)	•	•	2+	no
TCP-M	no	yes	(•)	•	•	1	no
EMIPv6	no	yes	(•)	•	•	1+(# of peers)	yes
TeleMIP	no	no	(•)	•	•	2+	no
CellularIP	yes	no	(•)	•	•	3+	yes
HAWAII	no	no	•	•	•	3+	no
mSCTP	no	yes			•	1+	no
SIP	no	yes/no	•	•	•	1+	yes (with state-full servers)
smartAR	yes/no	no		•		1+	yes

We should use SIP to solve the simultaneous movement problem. However, in this case we should employ SIP domains and SCTP stacks. Another approach is to exploit network component that are always present, like the AR. Our work has been inspired by similar works, for instance DHMIP [11]. However, in DHMIP, the details of protocol signalling are not shown and HA are still present. Moreover, the case of simultaneous movements is not faced. We suppose that in every domain the $AR(\cdot)_{dk,i}$, $k = 1,2$, is smart, that is it can store for a certain period of time the new location of the MN that just left the subnetwork. For example, we can suppose that the AR can store a cache whose values are 4-tuples. Every tuple is in the form $(MN_a, CoA, NextHop, F)$, where MN_a is the address of the MN, e.g. its MAC address, CoA is the CoA associated with MN_a in a subnetwork, $NextHop$ is the AR to which packets should be forwarded, and F is a state flag for the CoA. Initially, at the very first association of the MN with an AP of a particular AR, the cache state flag is set to "P", i.e. "Preferred". Whenever the MN leaves the subnetwork $AR(\cdot)_{dk,i}$, its cached values will change. For instance, the state flag will be set to "D", i.e. "Deprecated", because the MN is no longer reachable to that address, or rather that address is no longer present in that subnetwork[4]. For this reason, we need some extra signalling messages. In particular, we need to know when the MN addresses are to be considered as deprecated into the cache of AR. Moreover, we need to provide every AR with the address of the next hop whenever the address flag is set to "D". Even if we do not face here the comparison with current available technologies, we suppose that ARs update their own caches by means of special L3 messages we name virtual Binding Update messages (vBU). The vBU carries the information of a cache tuple. By this way, the $AR(\cdot)_{dk,i}$ can forward packets whose destination is "Deprecated" to the proper "Preferred" address, e.g. from $AR(\cdot)_{dk,i}$ to $AR(\cdot)_{dk,i+1}$. Let us note that in this scheme, while the MN is changing subnetwork, the CN uses the very

---

[4]This mechanism is similar to the addressing scheme of IPv6.

1 Algorithm of smartAR

```
# i: address of the present AR
# H: cache of address values
# я = Hᵢ(x): return the tuple which match x in H
# Aᵢ, components of я
for each packet P do
    я = Hᵢ(dest(P))
    if A₄ ="D" and A₃==i:  per flow buffering
    else:  forward to A₃
    end
    if ∃я = Hᵢ(MNₐ): A₂ = CoA of P
end
for each vBU do
    я = Hᵢ(vBU),  A₄ ="P", A₃ =source of vBU
end
for each "syslog" do
я = Hᵢ(MNₐ)
    if Associated:
    if я = 0:  create tuple (MNₐ,·,i,D) in H
    else:  A₃ = i
    send vBU
    end
    if DeAuthenticated:  я = Hᵢ(MNₐ), A₄ = D
end
```

first address retrieved by mean of f, i.e. the CoA_i in Fig. 7. With this new AR in mind, we make the following observations. First of all, the signalling messages of the u function are local messages, they are exchanged among ARs which usually are geographically close one another. The g function is executed inside the AR, and, potentially, the location server are no longer needed, until we won't furnish a centralised naming service. Moreover, in order to change the state flag inside the AR cache, we need special signalling messages both for the hand-in and hand-off of every MN. This is not a hard task, because commercial APs can send dedicated log messages, called "syslog". These messages can report on the time instants of association and deauthentication to/from an AP. In this work, for simplicity we assume that every AR has one AP. The pseudo-code of the operations executed within the smartAR is given in Algorithm 1. We are assuming that the vBU message has been sending towards the correct AR, in particular the old AR the MN belonged to. Therefore, the address of the old AR should be embedded into the link layer frame. Alternatively, sending the vBU to the nearest AR will result in a simpler but effective technique. Another of that Algorithm is that the smartAR must know the CoA associated with a particular MAC address, otherwise the forwarding will be impossible. We suppose that mechanisms similar to [7] are available. It is worth noting that the smartAR is always aware of the most updated address of the MN. Accordingly, the smartAR can indeed handle both single and simultaneous movements scenario. In the following section, we compare the above solution by means of a cost metric.

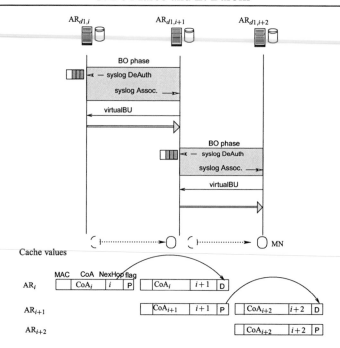

Figure 7: Basic operations of smartAR.

6   Performance Analysis

In this section, we compare the hand-off latencies of smartAR with other IP micro-mobility protocols. This solution can be compared with e2e solutions only, for example SIP and SCTP. In this section, we ignore other e2e solutions as [5,6], because they use a P2P system which is deeply different from common client-server mechanisms. To quantify pros and cons of the protocols, we introduce a cost for each function described in Section 2. We use a methodology similar to [11]. The cost functions are as follows.

- c_f is the cost of location retrieval of the correspondent MN/CN;

- c_g is the cost to send a packet towards the MN/CN;

- c_u is the cost of location update.

For example, in SIP, we have that c_f depends on the network delay between the CN and the SIP server; the cost c_g is a function of the delays between the SIP server and the foreign network SIP proxy, and the forwarding delay between two end peers. The cost c_u depends on the delays of location update whenever a hand-off takes place. It is clear that the overall cost function, i.e. the cost caused by location updating and forwarding, depends also on the arrival rate of packets and the rate of hand-off [4, 11]. Before going ahead, we summarise the symbols we use.

- C_s and C_m are the overall cost functions for the single movement and simultaneous movement scenario, respectively.

- $p(i)$ is the probability of crossing i subnets between two consecutive packet arrivals.

- λ_p is the rate of packet arrivals and λ_h is the rate of hand-off which depends on the residence time in a particular subnetworks.

- ρ is the Call Mobility Ratio (CMR), that is $\rho = \frac{\lambda_p}{\lambda_h}$.

6.1   SIP

If we use SIP, for the single movements case, the c_f function is due to the Re-INVITE message which is aimed to change the identity information at the correspondent peer. While the c_u is the cost of reliably sending a REGISTER message towards the SIP server. We have:

$$C_s(\rho) = \sum_{i=0}^{\infty} ip(i)(c_u + c_f) + c_g. \tag{3}$$

The function c_g adds the delays due to IP network. For instance, in MIP this cost would be greater. It can be proved [11] that:

$$C_s(\rho) = \frac{c_u + c_f}{\rho} + c_g. \tag{4}$$

This formula is valid for any distribution of residence time inside a subnetwork. If we used MIP, c_g would count the cost of sending a packet from the HA towards the MN. For the simultaneous movement case, we must consider the probability of simultaneous movement, we call p_s. The parameters which affect this probability are the λ_h and the network delays between MN and MN, SIP server and MNs. Note that we are assuming that the hand-off rate is the rate of crossing a subnetwork. If we consider only L3 hand-off, the BO phase is affected only by the MN address configuration delays. We have:

$$C_m(\rho) = 2C_s(\rho)p_s + (1 - p_s)C_s(\rho) = C_s(\rho)(1 + p_s). \tag{5}$$

If $p_s = 0$ we have the single movement case, $C_m = C_s$, otherwise we must count two times the overall cost function, one for each MN. If the network paths are not symmetric, we should distinguish the overall cost function according to the MN they refer to. For example, if we have MN_1 and MN_2, we can refine Eq. 5 as:

$$C_m(\rho) \quad = \quad p_s\left(C_s^{MN_1} + C_s^{MN_2}\right) + (1 - p_s)\max\left(C_s^{MN_1}, C_s^{MN_2}\right), \tag{6}$$

and

$$
\begin{aligned}
C_s^{MN_1} &= C_s(\rho_1) \\
C_s^{MN_1} &= C_s(\rho_2),
\end{aligned} \tag{7}
$$

by inherently assuming that c_u and c_f are different between $C_s^{MN_1}$ and $C_s^{MN_2}$.

6.2   SCTP

In this case, we can compute C_s only, because as said before SCTP does not fit the requirements for simultaneous movements. According to the network topology, the function c_u and c_g could vary between two solutions. Without introducing further symbols, we simply compare these functions in the Section 7, first for SIP and then for SCTP.

Table 2: Cost functions parameters.

	c_u	c_f	c_g
SIP	$d_{\text{M-S}} + d_{\text{S-M}}$	$d_{\text{M-S}} + d_{\text{S-M}}$	d_{e2e}
SCTP	d_{e2e}		d_{e2e}
smartAR	$c_u^{\text{vBU}} = d_{\text{AR}}$		$c_g^{\text{AR}} = d_{\text{AR}},$ $c_g = d_{e2e}$

6.3 smartAR

If we use smartAR as described in Section. 5, we stress that the updating function is no longer necessary, or rather the registrations of new CoA back to the location point, either SIP server or CN, are not strictly necessary. In fact, the vBU message acts as a "hidden" registration and the smartAR cache acts as temporary location server. We then indicate by c_u^{vBU} the cost of updating MN locations among smartARs. On the other hand, now the packets towards the MN are forwarded through k ARs, where k is the number of AR traversed by the MN. We name c_g^{AR} the cost of this local forwarding. For simplicity, we do not bound k. Then, we have, for the single movement case:

$$C_s(\rho) = \sum_{k=0}^{+\infty} p(k)k\left(c_u^{\text{vBU}} + c_g^{\text{AR}}\right) + c_g = \frac{c_u^{\text{vBU}} + c_g^{\text{AR}}}{\rho} + c_g. \tag{8}$$

For the simultaneous movement case, as in Eq. 5 we have:

$$C_m(\rho) = \left(\frac{c_u^{\text{vBU}} + c_g^{\text{AR}}}{\rho} + c_g\right)(1 + p_s). \tag{9}$$

7 Numerical Results

In order to compare the above solutions, we perform some basic numerical simulations. It is straightforward that the cost metrics depends on the network topology. We assume that ARs are co-located, as this is a reasonable hypothesis. We introduce the following parameters:

- d_{AR} which is the transmission delay of a vBU between two ARs (in the same domain).

- $d_{\text{M-S}}$ is the transmission delay between the MN and the SIP server.

- d_{e2e} is the end-to-end delay between two peers, e.g. two MNs.

- $d_{\text{S-M}}$ is the delay between the SIP sever and the MN/CN.

We are also supposing that the transmission cost of a packet from the AR towards the MN and vice versa is fixed, then it can be neglected in the computation. For instance, d_{e2e} then represents also the transmission delay between two ARs. For the range of parameters we chose the values in arbitrary time scale unit, as shown in Table 3.

Table 3: Delays values for simulation.

d_{AR}	$d_{M\text{-}S}$	d_{e2e}	$d_{S\text{-}M}$
1	10	$1 \div 100$	100

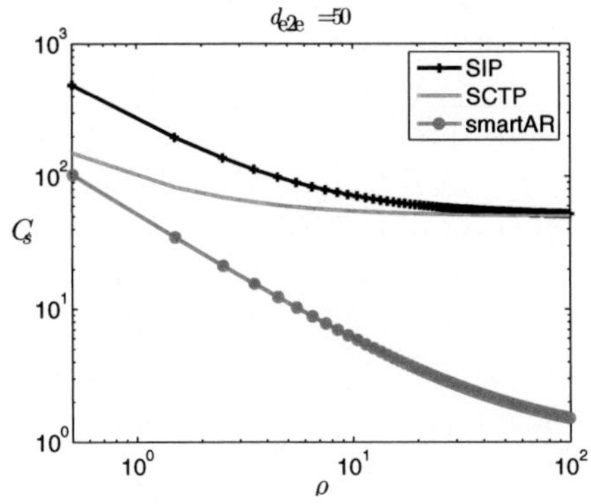

(a) Single movement, low e2e delay

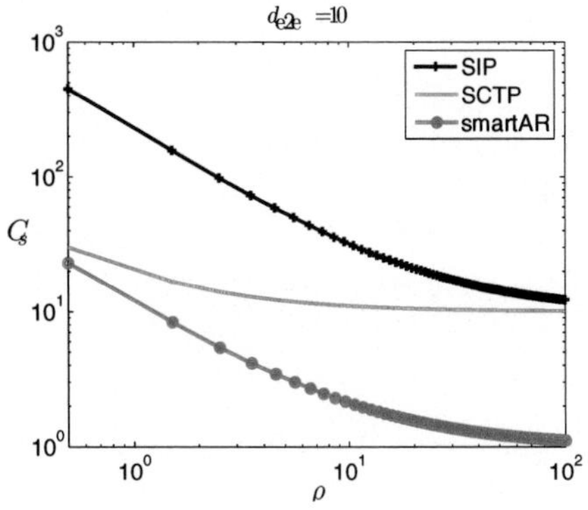

(b) Single movemen, high e2e delay

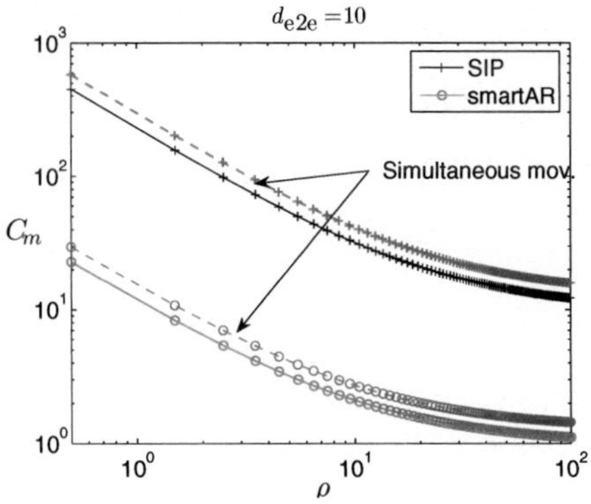

(c) Simultaneous movement case

Figure 8: Hand-off cost functions comparison.

In Fig. 8, we show the plot of the cost functions. Low values of ρ mean a high mobility grade. For high ρ, we obtain that SIP and SCTP are comparable, because the dominant term is c_g. For low mobility grade, SCTP outperforms SIP because it updates the localisation information in end-to-end fashion. Thus, it is clear that as long as the SIP server is *far* from the MN, the SIP solution introduce the additional delay due to the distant SIP server. The smartAR outperforms both SIP and SCTP, because the updating functions have been executed in *local*. Moreover, this behaviour holds also for simultaneous movements, as shown in Fig. 8-c. The only case for which both SIP and smartAR results in the same cost value is when the SIP server is co-located, that is it is topologically and physically placed in the same subnetwork of the ARs.

8 Conclusion

In this chapter, we gave a classification of the current IP micro-mobility problems based on three simple mobility functions. By means of a general taxonomy, we recognised that some combinations of standard mobility protocols, like MIP+SIP, introduce redundancy in the network. SIP is a well understood application protocol, while MIP and its variants are network protocols. However, both require the deployment of additional network components. With a kind of criticism, we can say that if standard applications, like Web and FTP, are not supported by SIP and/or MIP based domains, the micro-mobility can not be realised. Other distributed solutions are based on the P2P paradigm which for years has been considered for file sharing only. Its scalability and robustness have been attracting the scientific community as well as network architects to solve the problem of localisation of MNs. However, the real benefits of such architecture in the case of micro-mobility with fast users have not been yet analysed. On the other hand, transport layer solutions to the micro-mobility problems are possible whenever the protocol permits the dynamic change of the destination IP address, like SCTP with ASCONF messages. The first attempt to end-to-end mobility was the TCP-Migrate protocol [20], which used the updating of DNS to control the location of the MN. Here, we saw that SCTP with the ASCONF messages support can quickly handle the updating function. Also SIP can be thought as an e2e protocol for micro-mobility. In fact, the Re-INVITE message does not strictly need to be routed through a SIP server. However, simultaneous movements case is a harmful case for e2e solutions which are not able to correctly execute the updating and/or forwarding function. Here, we showed another solution, based on an enhanced version of the AR, we named smartAR. This solution borrows some concepts of FMIPv6 for the fast hand-off, but without introducing further messages than the vBU message. As a matter for further investigation, we plan to build a complete architecture, along with performance evaluation of the smartAR in a real test-bed. Eventually, the case of different CMRs and the exact relation between p_s and ρ are to be studied in further works.

References

[1] P. D. Cleyn, N. V. den Wijngaert, L. Cerdà, and C. Blondia, *Computer Networks*, **45**,3, 345–361 (2004).

[2] S. Das, A. Misra, and P. Agrawal, *IEEE Personal Communications*, **7**, 4, 50–58 (2000).

[3] G. De Marco, S. Loreto, and L. Barolli, *LNCS*, Springer-Verlag (2005), Performance analysis of ip micro-mobility protocols in single and simultaneous movements scenario, In Proceedings of UISW'05, 3823, December 2005 (Nagasaki, Japan) pp. 443–451.

[4] Y. Fang and W. Ma. *Mobility management for wireless networks: modeling and analysis*. Wireless Communications Systems and Networking. Kluwer Academic/Plenum Publishers, (2004).

[5] C. Guo, H. Wu, K. Tan, Q. Zhang, J. Song, J. Zhou, C. Huitema, and W. Zhu, End-system-based mobility support in ipv6, *IEEE Journal of Selected Areas in Communications (JSAC)*, **23**, 1, 2104–2117, (2005).

[6] C. Guo, Q. Zhang, and W. Zhu, A seamless and proactive end-to-end mobility solution for roaming across heterogeneous wireless networks, *IEEE Journal on Selected Areas in Communications*, **22**, 5, 834–848, (2004).

[7] S.-H. Hwang, Y.H.Han, C.-S. Hwang, and S.-G. Min, *LNCS*, Springer-Verlag (2004), An address configuration and confirmation scheme for seamless mobility support in ipv6 network, In Proceedings of Wired/Wireless Internet Communications Conference, 2957, pages 74–86.

[8] IETF. *Session Initiation Protocol (SIP) Session Mobility*, February 2005.

[9] S. Koh and Q. Xie, *M-Sctp with Mobile Ip for Transport Layer Mobility*, IETF - draft (2004).

[10] R. Kooli, *Fast Handovers for Mobile IPv6*, IETF RFC 4068, (July 2005).

[11] W. Ma and Y. Fang, *IEEE Journal on Selected Areas in Communications*, **22**, 4, 664–676, (2004).

[12] P. McCann, *Mobile IPv6 Fast Handovers for 802.11 Networks*, IETF RFC4260, (November 2005).

[13] C. Perkins and J. Arkko, Mobility support in ipv6, IETF RFC3775, (June 2004).

[14] I. Ramani and S. Savage. Synscan: Practical fast handoff for 802.11 infrastructure networks. In Proceedings of INFOCOM'05, March 2005, pp. 675–684.

[15] P. Reinbold and O. Bonaventure, *IEEE Communications Surveys & Tutorials*, **5**, 1, 40–56 (2003).

[16] M. Riegel and M. Tuexen, *Mobile sctp*. IETF - draft.

[17] J. Rosenberg, H. Schulzrinne, G. Camarillo, A. Johnston, J. Peterson, R. Sparks, M. Handley, and E. Schooler, *SIP: Session Initiation Protocol*, IETF RFC 3261, (June 2002).

[18] M. Shin and A. W. Arbaugh, Improving the latency of 802.11 hand-offs using neighbor graphs. In Proccedings of ACM MobiSys'04, June 2004, pp. 70–82.

[19] Skype, Inc. www.skype.com.

[20] A. C. Snoeren and H. Balakrishnan, An end-to-end approach to host mobility, In Proceedings of the 6th IEEE/ACM International Conference on Mobile Computing and Networking (MobiCom), August 2000, pp. 155 – 166

[21] H. Soliman, Flarion, C. Castelluccia, K. E. Malki, and L. Bellier, *Hierarchical Mobile IPv6 Mobility Management (HMIPv6)*, IETF RFC 4140, Experimental, August 2005,

[22] R. Stewart et al, Stream Control Transmission Protocol (SCTP) Dynamic Address Reconfiguration, IETF - draft, (December 2004).

[23] A. G. Valkó, *SIGCOMM Comput. Commun. Rev.*, **29**, 1, 50–65 (1999).

[24] K. Wong, A. Dutta, K. Young, and H. Shulzrinne, Managing simultaneous mobility of ip hosts, In *Proceedings of MILCOM'03*, 22, October 2003, pp. 785–790.

In: Mobile Multimedia: Communication Engineering …
Editors: I.K. Ibrahim and D. Taniar pp. 63-74
ISBN: 1-60021-207-7
© 2006 Nova Science Publishers, Inc.

Chapter 4

MOBILE DATA COMMUNICATION FOR EMERGENCY RESPONSE

Andreas Meissner and Zhou Wang

Fraunhofer Integrated Publication and Information Systems Institute,
Darmstadt, Germany

Abstract

This chapter looks into a new and rather challenging application domain for mobile data communication: emergency response operations, carried out by fire departments, police, and other emergency services. Following a review of mobile network engineering and current technology, fundamental requirements imposed by this application domain on mobile information systems are identified. As a case study, Fraunhofer's MIKoBOS system is presented, an integrated mobile information and communication system for emergency response operations that allows for reliable data communication and information exchange within an emergency site as well as between a site and the headquarters of organizations involved. It supports multiple communication networks at both local and wide area level and provides responsible personnel with anytime-anywhere access to relevant information.

Introduction

Mobility is becoming one of the distinguishing marks of today's information society. It is of ever increasing interest to be able to access networked information and services from portable devices such as laptops, PDAs, and even mobile phones, at any time from anywhere. The wide deployment of easily accessible wireless and mobile communication networks is now making mobile information access a reality. A wide range of wireless network technologies, such as WLAN, GSM/GPRS, UMTS, and Bluetooth, can be used for mobile data communication. Nowadays, many mobile devices are equipped with multiple wireless technologies. For example, a laptop may have LAN, WLAN, Bluetooth, and even GPRS interfaces, and a PDA or smart phone may have GSM/GPRS, WLAN and Bluetooth interfaces. Considering that wireless networks vary greatly regarding availability, bandwidth, delay, response time, error rate, network coverage, and cost, it has been recognized that no single access technique can fulfill all the requirements brought along in a scenario as difficult and complex as an emergency response operation [9], an application domain this chapter will be centered around. Moreover, user mobility means continuous changing of location and environment. These highly dynamic factors in wireless networks and the mobility of users have led to the demand for seamless handoff between diverse wireless networks, so that various networking resources can be utilized in a collaborative and complementary manner.

In a strong contrast to the achievements of mobile communication technology, and despite their inherently mobile operations, fire and police departments as well as other emergency services in many countries still rely on manual information exchange and processing procedures, with modern information technology being deployed only in their headquarters, if at all [2] [7] [8]. Even today, it is very common for fire departments to carry files of paper in their command post vehicles and to use voice radio for the exchange of information between an emergency site and their headquarters. However, as an emergency situation may develop very dynamically, it is vital to have the right information at the right time, at the right place. The more this is accomplished, the easier it is to overcome what is known as the "chaos phase" at the beginning of a response operation, and to master the situation while it is evolving. Obviously, this calls for mobile data communication technology and for mobile information systems built on top of such networks.

The few commercial solutions that are currently available range from small special-purpose notebooks to the replication of an office-style environment in the command post vehicle (with email-based information exchange accomplished through a hook-up with the wired ISDN network), or even the installation of a full-blown C3I client in the vehicle that provides the same work environment and functionality a dispatcher at the headquarters would be used to. Given the fact that emergency responders face a difficult, sometimes sketchy data communication environment, and that the situation often does not allow for complex user interaction, and that integration with existing headquarters information systems is a key for acceptance, it is evident that the solutions in the market still leave much room for improvement. Moreover, most of these solutions do not even address the challenge of providing support for senior or frontline personnel roaming at the site away from their vehicle, thus requiring a wirelessly connected personal information device.

This observation, along with the awareness of state-of-the-art of mobile communication technology, prompted Germany's Fraunhofer Gesellschaft to design MIKoBOS, an easy-to-use Mobile Information and Communication System for Public Safety Organizations. MIKoBOS contains both a command post vehicle client that connects to an existing dispatch information system in the headquarters (e.g. the commercial CKS-112 system [17]) and a component designed for PDA-style devices in the sphere of staff roaming at the site. The basic idea is that selected static and dynamic information at the headquarters may be accessed in a downstream manner from the emergency site, and that some information – such as situation reports – may be created at the site and transferred upstream directly into the headquarter system. Regarding communication means, MIKoBOS uses a variety of wide area communication technologies, including satellite systems, for data exchange between the site and the headquarters, and it uses WLAN for site coverage.

This chapter provides a case study snapshot of MIKoBOS at its prototype stage. In the following sections, mobile network engineering is discussed as a primer, and requirements imposed on the system by the emergency management application domain are discussed.

Next, a detailed and specific look is taken at the MIKoBOS architecture, including system and network aspects. Finally, concluding remarks sum up the chapter.

Mobile Network Engineering

Although multiple network interfaces on mobile devices allow them to attach to multiple heterogeneous wireless networks, it remains a challenging issue to enable seamless handover between diverse access networks. The issues related to vertical handoff include the selection of the network link, the switch from one network link to another, and the provision of connectivity awareness to applications and users. Each connection link has its own characteristics by nature, with regard to availability, data rate, coverage, and cost – the selection of the link is therefore highly dependent on user preference, network availability and application requirements. For example, cost might be an important factor for private consumers, network security and bandwidth might be critical for business people, while high availability and reliability have to be taken into consideration for emergency services. Moreover, applications often need to be aware of network status, such as connected/ disconnected state, available bandwidth, or cost, so that they may adapt their behavior accordingly. Therefore, some kind of interfaces are desired to let applications and users query network information.

Various solutions have been proposed at different layers. While cellular networks, e.g. GSM/GPRS/UMTS, and WLAN provide link layer handover within the same access network, they do not support handover between different network technologies. Currently 3GPP is working on the architecture for WLAN and 3G system interworking [23]. At the network layer, Mobile IP [3] [10] provides transparent mobility support by tunneling all traffic to the mobile host through a home agent at its home network. However, Mobile IP is a pure routing solution and suffers from inefficient routing, comparably long handoff latencies, and sometimes conflicts with network security solutions. Several improvements, such as Cellular IP [16], Hierarchical Mobile IPv6 [13] have been suggested to address these issues. At the transport layer, regular TCP uses the IP address to offer services and distinguish between connections on well known ports, and thus does not support IP address change during an established TCP connection. Some extensions of TCP have been proposed to support vertical handover by allowing seamless migration of IP address changes. Examples are TCP-Migrate [12], TCP-MH [6], MSOCKS [5]. Similarly, mobile SCTP [11] is a mobile extension of the SCTP protocol [14]. However, both network layer and transport layer based extensions rely on some kind of new network infrastructure component such as the home agent, or some change of existing network protocols, which makes their deployment difficult.

On the other side, several efforts have been made to support vertical handover on end hosts. Windows Mobile 2005 provides a Connection Manager for centralizing and automating the establishment and management of various kinds of network connections [22]. When one or more applications request a connection, they only need to specify a connection name and network name. The Connection Manager establishes the connection using the optimal connection path. The choice of path for a particular connection depends on cost, security, and specific network considerations for the application. In this way, users do not know which connection path is chosen and yet they can be assured to use the optimal path at all times. The Connection Manager also tracks which connections are in use or requested by applications. It closes unused connections and automatically disconnects when they are idle for a specified period of time. Although the Connection Manager handles many different types of connections, it does not support handover between different networks yet. Other application layer or middleware solutions include PACE and CMM. In PACE [1], the authors propose a concept of vertical handover supporting pervasive computing in future wireless networks. The aim is to enable a context-aware handover in heterogeneous networks. Their concept uses context information, such as user device, location, and network QoS, to decide and perform a redirection of the communication streams between multiple network interfaces. The decision of a handover can be influenced by an application initialization or device change, the moving out of networks or entering of new networks, and the network's QoS changes. The selection process considers certain factors, e.g. user device preferences, bandwidth, jitter, delay, packet loss, and bandwidth fluctuations. However, the deployment of PACE needs some additional components such as context repository and adaptability manager in the infrastructure. Another middleware architecture for connectivity management is CMM [15]. CMM is built upon the network and transport layers. CMM provides flexible support to adaptive network applications with a set of functions and interfaces, including both connectivity context awareness and adaptive connectivity management. Through a connection monitor, it provides interfaces for applications to query information about available network links. It also provides interfaces for applications to subscribe to connection events and receive notification on their occurrence. The channel management agent is responsible for creating and maintaining connection channels, and it provides a policy-based mechanism to evaluate the best connection for each channel at a given moment. If a connection is temporarily unavailable, CMM can suspend the channel operations and wake it up later when a new connection becomes available.

Emergency Management Domain Survey

Following the survey of mobile network engineering and available technologies, particularly for vertical handover, this section addresses the domain requirements imposed on mobile information systems to be used in emergency management.

According to the definition of the U.S. Federal Emergency Management Agency [18], disaster management can be divided into four different phases, although they are not necessarily distinct: mitigation, preparedness, response and recovery. While the former two are mid to long term activities requiring information exchange in an office-like manner, response and, to a lesser extent, recovery are time critical activities requiring fast and reliable communication at the emergency site and between the emergency site and the different locations of involved organizations like police, fire departments, and medical emergency services.

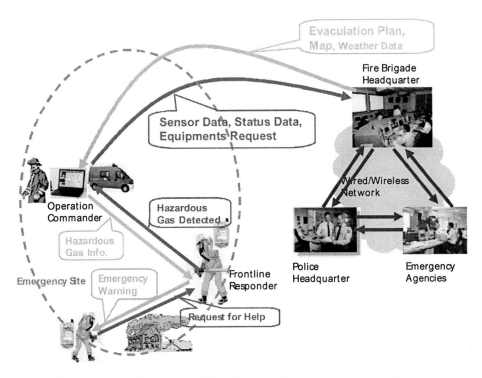

Figure 1. Sample Information Flow during an Emergency Response Operation

Figure 1 illustrates a typical information flow during the response phase of an operation, which involves frontline responders, operation commanders and headquarters for intra and inter organization coordination. Staff at the site need to have access to various distributed information sources, such as hazardous material databases or digital floor maps of buildings or technical infrastructure. They also need to receive from their headquarters data such as operation plans, resource allocation plans, and dynamic information such as weather updates or simulations of possibly hazardous situations, e.g. when smoke is moving towards an inhabited area. Selected information has to be forwarded proactively to the frontline responders. For such communication it is crucial to avoid information overload, i.e. the system must ensure that frontline responders receive only the necessary information exactly when it is needed. At the same time, operation staff should report current site status to the headquarters, such as sensor data or requests for additional resources, to allow for a common operational picture forming the basis for proper decisions at upper levels. Nowadays, the communication among and between frontline responders, operation commanders and headquarters is still accomplished mainly through traditional voice communication, which has several disadvantages in nature. The channel is usually insecure and error-prone. The frontline responder has to listen to voice messages carefully, at the same time he has to be focused on his mission. This situation calls for adopting modern information technology to improve communication and coordination during emergency response operations.

Based on interviews with experts in the field and a technology analysis, the following sub-sections briefly summarize some fundamental requirements for an information system for emergency response.

Provision and distribution of information. The description of the information flow above shows that, apart from a secure and reliable connectivity, it is important to guarantee that the right information be transferred at the right time to the right person. Some intelligent filtering and distribution mechanisms, primarily at the site control, are thus essential to avoid information overload for frontline responders. However, it should be guaranteed that all necessary information arrive in time. To this end, both push-based and pull-based information delivery approaches have to be supported. Moreover, the information should be distributed and presented in an appropriate format, since responders with different roles may have different types of terminals like laptops, PDAs, or even radio handsets, and user interaction may be difficult. An additional challenge is the integration with existing information systems at the headquarters. Such systems usually keep vast amounts of stored data, e.g. on hazardous materials, which might need to be accessed by on-site personnel. At the same time, staff at the headquarters, due to their physical distance to the emergency site, need up-to-date information from the site for the purpose of coordinating resources and taking decisions. The integration with existing headquarter information systems is essential for acceptance in practice.

Rapidly deployable reliable data communication. Digital data communication provides not only improved communication quality and security, but also the possibility for accessing remote databases, communicating with head offices by using video conferencing and other collaborative tools. However, considering the fact that the public fixed communication infrastructure at an emergency site is often unavailable, destroyed, or overloaded, communication for response operations should rely on some kind of infrastructure which can be deployed by responders in a short time after they reach the emergency site. This requires readiness for dynamic and automatic configuration and adaptation of a communication infrastructure. Besides rapid deployment, disaster relief requires reliable and robust communication that is available in virtually any situation at any time. The communication technology has to resist environmental strains like great heat or water, and it has to provide connectivity at all times [24]. While resisting the environmental strains is mainly a challenge for hardware developers, continuous connectivity calls for some level of redundancy in the topology of the network to cope with unexpected events, and, at the terminal side, support for multiple networks – as outlined above – proves advantageous.

Mobile Data Management. Mobile communication in a layered, fast changing network requires specific mechanisms to guarantee a consistent information exchange. It is a challenge to ensure data access and data consistency over low-bandwidth unreliable wireless links in an ever-changing environment. Mechanisms like caching or prefetching help optimizing bandwidth usage and reducing the impact of – expected or unexpected – disconnections. Moreover, considering that critical data must obtain priority if the communication channel deteriorates, effective bandwidth management is an essential part in mobile data management.

Current IT support for emergency response usually covers only a subset of these aspects. The conventional headquarters information systems are conceived for stationary environments and based on fixed network infrastructures. Apparently there are no integrated solutions yet which provide reliable ubiquitous access to information from the emergency site and meet the requirements mentioned above.

MIKoBOS Architecture

After a discussion of mobile network engineering and specific domain requirements in previous sections, this section brings technology and requirements together and provides a case study of MIKoBOS, a system designed to support emergency services in their response operations, making some of the technology outlined above available to them. The primary aim of MIKoBOS is to improve the communication and the coordination within and between public safety organizations during disaster and emergency response operations. Therefore, apart from meeting the requirements mentioned above, it is essential that the architecture of MIKoBOS be in line with the workflow of emergency operations. The next sub-sections present the architecture of MIKoBOS, starting with the overall system architecture and a description of the functional components. Network aspects are discussed subsequently.

System Architecture

The typical organizational structure of emergency services and the nature of emergency response operations imply that MIKoBOS is a multi-level distributed application. The MIKoBOS system consists of three application components: MIKoBOS-LS for headquarters, MIKoBOS-TEL for on-site operation commanders and MIKoBOS-EP for frontline responders roaming at the site. The software architecture of the overall MIKoBOS system is illustrated in Figure 2.

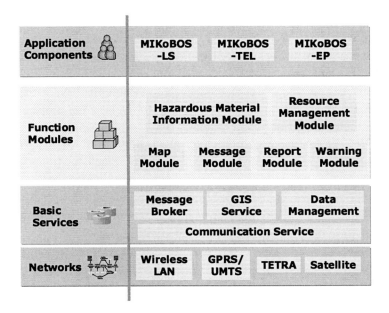

Figure 2. MIKoBOS System Architecture

In this architecture, the lowest layer builds the communication infrastructure for MIKoBOS. Its detailed structure will be discussed in the next sub-section. The basic services provide common functionalities for function modules. The communication service is a key component that is designed for the dual purpose of supporting multiple network technologies and providing a consistent interface for other modules to utilize communication regardless of the underlying technology used. To deal with the variability effect of underlying heterogeneous networks on communication performance, two application-level lightweight communication protocols are implemented by the communication service, based on TCP and UDP, respectively. Both protocols have been designed with the intention of overcoming disadvantages of unreliable wireless communication. The communication service is also responsible for alerts of communication status changes, managing bandwidth allocation, and adapting the data stream to available communication conditions according to associated policies like priority. Upon the basic services, different function modules, such as the hazardous material information module and the resource management module, are built.

The application components, i.e. MIKoBOS-LS, MIKoBOS-TEL, and MIKoBOS-EP, are configured to run on different hardware platforms with customized functionalities and user interfaces (Figure 3). MIKoBOS-LS is designed for dispatchers in headquarters and acts as a bridge for data exchange between MIKoBOS and the existing dispatch information system such as CKS-112 [17]. With MIKoBOS-LS, headquarters staff can be informed on the current status at the emergency site in near real time through means such as collected data, text message, and pictures, which assist them in making a direct situation evaluation in order to come up with more informed decisions and advice.

MIKoBOS-TEL, running on a robust notebook PC which is installed inside a command post vehicle equipped with power supply and diverse communication technologies, enables the operation commander at the emergency site to have access to central headquarters databases, to request new resources, and to report the current situation. All data exchange is scheduled by the communication service based on available communications, message priority and message size. At the same time, MIKoBOS-TEL allows the operation commander to communicate with frontline responders in the field. For example, values taken from portable sensors may be transferred from responders to the operation commander, where the data can be stored, aggregated, and forwarded to the headquarters if needed. Similarly, the operation commander can reach frontline responders in multiple ways, either by specifying receivers explicitly (e.g. point-to-point or broadcast) or by sending messages to a selected group based on content or geographic location.

While MIKoBOS-LS and MIKoBOS-TEL are mainly designed for operation control, MIKoBOS-EP, running on robust PDAs, is primarily conceived for frontline responders to send and receive information while they work away from vehicles. With MIKOBOS-EP, the frontline responders can access hazardous material information, transfer collected real time sensor data, send textual or visual situation reports to the operation commander, and receive instructions and warnings from the operation commander.

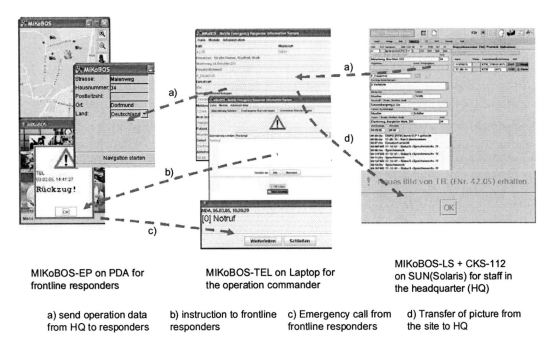

MIKoBOS-EP on PDA for frontline responders

MIKoBOS-TEL on Laptop for the operation commander

MIKoBOS-LS + CKS-112 on SUN(Solaris) for staff in the headquarter (HQ)

a) send operation data from HQ to responders

b) instruction to frontline responders

c) Emergency call from frontline responders

d) Transfer of picture from the site to HQ

Figure 3. MIKoBOS Applications

Network Architecture

The MIKoBOS network infrastructure is IP based and includes a local area network at the headquarters, wireless networks at the emergency site, and a variety of wireless wide area network links between the headquarters and the site, as shown in Figure 4. At the emergency site, broadband local wireless networks, such as IEEE 802.11 wireless LAN, are used. Three kinds of communication technologies are supported for the WAN connection: GSM/GPRS/UMTS-based public mobile networks, PSO-proprietary terrestrial trunked radio (TETRA) [21], and satellite communication. For all of them, clients are easy to set up and can be deployed in a short time. However, each of them has its advantages and drawbacks. The GSM-based networks are well known and widely used technology. They provide relatively wide bandwidth at affordable prices with coverage in almost all inhabited areas in many countries. However, their operation is heavily dependent on some kind of fixed infrastructure, such as base stations, which may be totally disrupted in large-scale disasters [4]. Another issue is that, due to its public accessibility, it is difficult to reliably give higher priority for emergency communication over non-emergency communication. Therefore, responders cannot rely solely on such public communication infrastructure in disaster response. In contrast to GSM/GPRS/UMTS, terrestrial trunked radio networks may be specially deployed for and exclusively used by public safety organizations, which ensures their availability to some extent. However, data communication is not the focus of the current TETRA (phase 1) technology. It supports only 28.8 kbps data transfer rate, in most cases even less. As another alternative for WAN connection, satellite communication is well known for its high availability. Satellite networks have very good outdoor coverage and are not affected by local disasters. New data oriented satellite services, such as Inmarsat's Regional Broadband Global

Area Network RBGAN [20] or Inmarsat ISDN MPDS (Mobile Packet Data Service) [19] make mobile satellite communication more attractive for public safety organizations. However, satellite communication, compared to the other technologies mentioned above, is expensive and, in case of geostationary satellites, usually needs an unobstructed line-of-sight to the satellite used. At the headquarters, a standard fixed connection to the Internet is configured for day-to-day usage with TETRA and satellite communication as backup channel.

Therefore, it can be concluded that there is no single communication technology that fulfills the above-mentioned requirements in emergency response. As a consequence, MIKoBOS uses a network architecture supporting multiple communication technologies simultaneously, which is an essence for emergency operations. By integrating various technologies into one platform and by using them flexibly and interchangeably, the various communication alternatives can complement each other. If one network is unavailable, the system can switch to use another available option. The switch between different technologies can be done automatically according to pre-defined policies, or manually by an operation commander. In this way, an acceptably high availability of communication can be guaranteed without incurring extensive unnecessary costs.

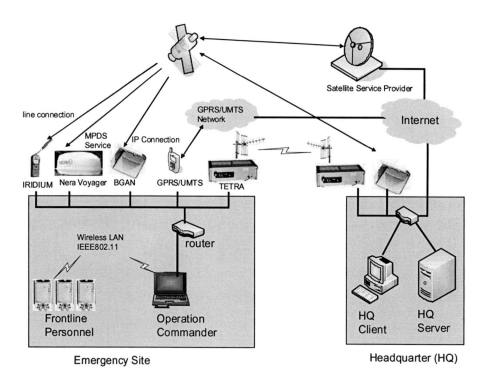

Figure 4. MIKoBOS Network Architecture

Experimental results on MIKoBOS' communication performance, including an analysis of its TCP vs. its UDP options and its responsiveness from the user's perspective with various underlying networks, may be found at [9]. An analysis of these results suggests that MIKoBOS may actually be used in practice. Inmarsat's RBGAN satellite network is especially appropriate, and given the fact that Inmarsat has initiated the upgrade of this service to BGAN with better global availability and even higher data rates, emergency services may even make it their primary choice for their wide area data communication needs in areas lacking a terrestrial infrastructure.

Conclusion

In this chapter, mobile network engineering technology has been reviewed, and a specific, highly challenging application domain for mobile data communication has been surveyed: emergency response operations. Bringing technology and application domain requirements together, Fraunhofer's MIKoBOS system has been presented as a case study of a mobile information and communication system for disaster and emergency response operations. From a technical point of view, the focus of MIKoBOS is to study the seamless integration of different types of mobile terminals and various heterogeneous mobile communication technologies for dynamic workflow management in the emergency response application scenario, ranging from wireless local area networks to wide area communication networks including GPRS/UMTS and advanced satellite communications. From the application's point of view, the contribution of MIKoBOS is to demonstrate the technical feasibility and potential to adopt mobile computing technologies in disaster and emergency operations. Experimental results show that MIKoBOS can actually be used even in difficult communication environments calling for a satellite link.

References

[1]   Balasubramaniam, S., and Indulska, J. (2003) Vertical Handover Supporting Pervasive Computing in Future Wireless Networks. *Computer Communication Journal*, Special Issue on 4G/Future Wireless networks. Vol 27/8, pp. 708-719, 2003

[2]   Grasse, T. (2005) Eine Systemarchitektur zur effizienten Steuerung von mobilen Einsatzkräften – Design und Implementierung. *Diploma Thesis*, FernUni Hagen / Fraunhofer IPSI, September 2005 (in German)

[3]   Johnson, D., Perkins, C., and Arkko, J. (2004) *Mobility Support in IPv6*. IETF RFC3775, June 2004, http://www.ietf.org/rfc/rfc3775.txt?number=3775

[4]   von Kirchbach, H. et al. (2002) *Bericht der unabhängigen Kommission der Sächsischen Staatsregierung zur Flutkatastrophe 2002*. (In German)

[5]   Maltz, D. A., and Bhagwat, P. (1998) MSOCKS: An Architecture for Transport Layer Mobility. INFOCOM '98, *Proceedings of Seventeenth Annual Joint Conference of the IEEE Computer and Communications Societies*, 1998

[6]   Matsumoto, A., Kozuka, M., Fujikawa, K., and Okabe Y. (2003) Tcp multi-home options. Technical report, draft-arifumi-tcp-mh00. txt, *IETF Internet draft*, October 2003

[7]   Meissner, A., Luckenbach, T., Risse, T., Kirste, T., and Kirchner, H. (2002) Design Challenges for an Integrated Disaster Management Communication and Information System. *The First IEEE Workshop on Disaster Recovery Networks*, June 2002, New York, USA

[8]   Meissner, A., and Steinebach, M. (2004) *Neue IT-Infrastrukturen im Notfall- und Rettungswesen - Potential und Risiko*. Kongress Netz- und Computersicherheit,

Universität Düsseldorf, Okt. 2003; Published by W. Bertelsmann Verlag in 2004, ISBN 3-7639-3205-4. (In German)

[9]   Meissner, A., Wang, Z., Putz, W., and Grimmer, J. (2006) MIKoBOS - A Mobile Information and Communication System for Emergency Response. *Proc. ISCRAM 2006, Third International Conference on Information Systems for Crisis Response and Management*, Newark/USA, May 14-17, 2006, Eds.: B. Van de Walle and M. Turoff, ISBN-13: 978-90-9020601-1

[10] Perkins, C. (2002) IP Mobility Support for IPv4. *IETF RFC3344*, August 2002, http://www.ietf.org/rfc/rfc3344.txt?number=3344

[11] Riegel, M., and Tuexen, M. (2005) Mobile SCTP. *IETF Internet Draft*, draft-riegel-tuexen-mobile-sctp-05.txt, 2005

[12] Snoeren, A. C., and Balakrishnan, H. (2000) An End-to-End Approach to Host Mobility. *Proc. of the Sixth Annual ACM/IEEE International Conference on Mobile Computing and Networking*, August 2000.

[13] Soliman, H., Castelluccia, C., El Malki, K., and Bellier, L. (2005) *Hierarchical Mobile IPv6 Mobility Management* (HMIPv6).
      http://www.ietf.org/rfc/rfc4140.txt?number=4140, 2005

[14] Stewart, R. et al. (2000) Stream Control Transmission Protocol, *IETF RFC2960*, http://www.ietf.org/rfc/rfc2960.txt?number=2960, 2000

[15] Sun, J., Riekki, J., Jurmu, M., and Sauvola, J., (2005) Adaptive Connectivity Management Middleware for Heterogeneous Wireless Networks, *IEEE Wireless Communications*, Dec. 2005

[16] Valkó, A. G. (1999) Cellular IP: a new approach to Internet host mobility, *ACM SIGCOMM Computer Communication Review*, v.29 n.1, January 1999

[17] Tyco Fire & Security, CKS-112: http://www.cks-systeme.de/cks_112.html

[18] U.S. Federal Emergency Management Agency FEMA: http://www.fema.gov

[19] Inmarsat ISDN MPDS Packet Data Service, http://www.inmarsat.com

[20] RBGAN, *Inmarsat Regional Broadband Global Area Network*, http://regionalbgan.inmarsat.com

[21] ETSI TETRA, EN 300 392-x, *Terrestrial Trunked Radio*, Voice plus Data (V+D)

[22] MSDN, Windows Mobile Version 5.0 SDK, Connection Manager,
      http://msdn.microsoft.com/library/default.asp?url=/library/en-us/mobilesdk5/html/mob5oriConnectionManagerWindowsMobile.asp

[23] 3GPP, Feasibility Study on 3GPP system to WLAN Interworking, *Technical Report*, 3GPP TR 22.934 v6.2.0, Sept. 2003

[24] Project MESA (2005) *Service Specification Group - Services and Applications; Statement of Requirements*, MESA TS 70.001 V3.1.2, Jan. 2005

In: Mobile Multimedia: Communication Engineering ...   ISBN 1-60021-207-7
Editor: I.K. Ibrahim and D. Taniar pp. 75-100   © 2006 Nova Science Publishers, Inc.

Chapter 5

TOWARDS COGNITIVE RADIO FOR EMERGENCY NETWORKS

Qiwei Zhang, Fokke W. Hoeksema,
*Andre B.J. Kokkeler, Gerard J.M. Smit**
Department of Electrical Engineering, Mathematics and Computer
Science, University of Twente, P.O. Box 217,
7500 AE Enschede, The Netherlands

Abstract

Large parts of the assigned spectrum is underutilized while the increasing number of wireless multimedia applications leads to spectrum scarcity. Cognitive Radio is an option to utilize non-used parts of the spectrum that actually are assigned to primary services. The benefits of Cognitive Radio are clear when used in emergency situations. Current emergency services rely much on the public networks. This is not reliable in emergency situations, where the public networks can get overloaded. The major limitation of emergency networks is spectrum scarcity, since multimedia data in the emergency network needs a lot of radio resources. The idea of applying Cognitive Radio to the emergency network is to alleviate this spectrum shortage problem by dynamically accessing free spectrum resources. Cognitive Radio is able to work in different frequency bands and various wireless channels and supports multimedia services such as voice, data and video. A reconfigurable radio architecture is proposed to enable the evolution from the traditional software defined radio to Cognitive Radio.

Keywords: Cognitive Radio, Emergency Networks, Spectrum Sensing, OFDM, Reconfigurable Platform

1 Introduction and Background

Recent studies show that most of the assigned spectrum is underutilized. On the other hand, the increasing number of wireless multimedia applications leads to a spectrum scarcity. Cognitive Radio ([1], [2]) is proposed as a promising technology to solve the imbalance

---

*E-mail address: G.J.M.Smit@utwente.nl

between spectrum scarcity and spectrum underutilization. In Cognitive Radio, spectrum sensing is done in order to locate the unused spectrum segments in a targeted spectrum pool and use these segments optimally without harmful interference to the licensed user. A lot of research projects have started to prove the concept of Cognitive Radio and to demonstrate its feasibility. In the Adaptive Ad-hoc Freeband (AAF) [3] project, we design a Cognitive Radio based wireless ad-hoc network for emergency situations. Although the AAF project addresses Cognitive Radio in a complete fashion from physical layer to networking issues, this book chapter mainly focuses on the physical layer related issues including: spectrum sensing, baseband transmission and physical layer reconfigurability.

In section one, we give background information on the AAF project followed by an introduction to Cognitive Radio and related research projects. Section two is dedicated to spectrum sensing which is an essential functionality of Cognitive Radio. An OFDM based Cognitive Radio system is discussed in section three. In section four, we propose a reconfigurable platform to support Cognitive Radio.

1.1   Requirements for Emergency Networks

Current day emergency services rely for data communications on public radio networks like GPRS. Sometimes in disaster situations, even GSM is used for voice communication between relief workers. However, in case of emergency the public networks may get overloaded. So, the use of generally available public networks is not considered to be reliable enough for emergency situations. Moreover, GSM/GPRS is an infrastructure based network and therefore highly susceptible to disasters in small and medium sized urban areas. A real-life example is the S.E. Fireworks disaster in Enschede, the Netherlands (May 2000), where a fireworks depot exploded and destroyed a large part of the city, killing 23 people and injuring more than a thousand. In the first hectic hours after the last of the large explosions, the municipal disaster command in particular had great difficulty to gain an overview of the extent of the disaster, and of the situation in the disaster area. Fire brigade, police and relief workers in the medical chain experienced severe communication problems, both internally and with each ther, because transmission equipment appeared not to be working, or was functioning inadequately.

In such a scenario it is vital that communication between relief groups is established as quickly and as easily as possible. The fact that the infrastructure-based networks in such an area may be destroyed raises the need to communicate using infrastructure-less wireless technologies. So, a Radio Access Network (RAN) or some type of cellular network (GSM, UMTS) will not do. Access to the core telecommunications network will only be available at the periphery of the disaster area where the damage to the infrastructure is less severe. The area of operation of the emergency network is confined to the disaster area, which is, for example, no larger than a circle with a 2.5 kilometer radius (comparable to the Enschede case). An ad-hoc networking approach will allow the relief groups to enter the disaster area and communicate with each other quickly. With slightly more effort, the wireless ad-hoc network can be hooked up to the original core telecommunications network almost instantly. The relief network must be able to handle multimedia traffic service to its users. A reliable real-time voice communication is especially important in emergency situations. Data messages like location coordinates, sensor data and even the health status

Table 1: Radio parameters of the TETRA System.

Parameter	Value
Frequency Range	$380 - 400$ MHz
Channel Spacing	25 kHz
TX Power (mobile)	$1 - 3$ W
Channel Access method	TDMA
Modulation method	$\pi/4$-DQPSK
Channel bit rate	36 kbit/s
Maximum data rate (gross bit rate)	28.8 kbit/s
Net data rate, non-protected:	n x 7.2 kbit/s
Low-protected	n x 4.8 kbit/s
High-protected	n x 2.4 kbit/s
	(n = 1, 2, 3 or 4)
Speech coding	A-CELP, 4.567 kbit/s
Range Rural	± 14 km
Range Suburban	± 4.5 km

of the rescue workers are also needed. Pictures and video can be used for surveillance and locating objects. To support multimedia services, different constraints on QoS have to be met. Real-time voice and video are sensitive to delay and jitter while data messages have high requirements on data loss and error rates. The voice and video provide more or less predictable network load due to their streaming behavior but data traffic is rather bursty in nature. In short, the network must be able to handle a wide variety of multimedia signals and has to deal with large, possibly unpredictable amounts of data. The large amounts of data require a large amount of radio resources.

There are also public safety networks like C2000 which is based on TETRA (see table 1).

One may observe that there are limited data communication facilities available in C2000. Also this network is infrastructure based (there are fixed base stations) and thus susceptible to the type of disasters we conceive.

The state-of-the-art systems lack capabilities (e.g. in offered data rates or multimedia traffic support) and are not disaster proof. The other major drawback of current emergency networks is their spectrum scarcity [4] because current emergency networks are assigned with a limited spectrum and fixed bandwith. As we mentioned earlier, the large amounts of multimedia data in the emergency networks require a large amount of radio resources. One band can easily get congested due to heavy traffic, which makes it inadequate for emergency use. If several fragmented bands are assigned to emergency use, the interoperability and the lack of standards will become another problem [5]. Therefore to alleviate this spectrum shortage problem, a radio which dynamically accesses free spectrum resources turns out to be an interesting solution. This approach is called Cognitive Radio according to Mitola [1], who first came up with the concept. He defined Cognitive Radio as a radio that can change its transmission based on interaction with its environment. This definition can be very broad, including knowledge of services, user behaviour and spectral usage. The

AAF project focuses on two key elements, i.e. searching for under-utilised spectrum, and (rapidly) adapting transmission settings accordingly.

The first step in the project is to focus on identification of free resources in the frequency domain by spectrum sensing. Subsequentially we identify an OFDM-based system which is theoretically optimal in approaching the Shannon capacity in the segmented spectrum by sending at different rates and powers over each subcarrier. We believe that the capacity to nullify individual carriers poses interesting opportunities for cognition, as was also observed in [6]. Cognitive Radio has to operate in different frequency bands, combat various negative effects of wireless channels and support various multimedia services. Therefore Cognitive Radio needs physical layer reconfigurability. So, we propose a heterogeneous reconfigurable platform to enable the evolution from the traditional software defined radio to Cognitive Radio.

1.2   Cognitive Radio

The idea of Cognitive Radio was first presented by Joseph Mitola III in his paper [7], where he proposed that Cognitive Radio can enhance the personal wireless service by a radio knowledge representation language (RKRL). This language represents knowledge of radio at all aspects from transmission to application scenarios in such a way that automated reasoning about the needs of the user is supported. Explained in a simple way, Cognitive Radio is able to autonomously observe and learn the radio environment, generate plans and even correct mistakes. A comprehensive conceptual architecture of Cognitive Radio was later presented in [1], where Cognitive Radio was thought as a final point of the software-defined radio platform evolution: a fully reconfigurable radio that changes its communication functions depending on network and/or user demands. Mitola's work covered interesting research subjects in multiple disciplines such as wireless communication, computer science and cognitive science.

On the other hand, the radio regulatory bodies in various countries found that most of the radio frequency spectrum was inefficiently utilized. In November 2002, the Federal Communications Commission (FCC) in the United States released a report [8] aimed at improving the management of spectrum resources in the US. The report concluded that the current spectrum scarcity problem is largely due to the strict regulation on spectrum access. The spectrum measurements conducted by the FCC indicated that only small portions of the spectrum are heavily used while other frequency bands are either partially used or unoccupied most of the time. So, spectrum utilization can be improved by making it possible for a secondary user (who doesn't have the license for spectrum) to access the spectrum which is not occupied by the licensed user (primary user). This secondary user has the awareness of the spectrum and adapts its transmission accordingly on a non-interference basis. This spectrum access and awareness scheme is referred to as Cognitive Radio by the FCC, which is a narrower definition compared with the original concept brought up by Mitola.

1.3   Projects Related to Cognitive Radio

Since the concept of Cognitive Radio first appeared, research activities have started around the world. Recently, Cognitive Radio became a very hot topic due to its impact on future spectrum policy which could fundamentally change the current status of radio communication. We mention a few research projects related to Cognitive Radio.

In Berkeley Wireless Research Center (BWRC), a dedicated Cognitive Radio Research (CRR) project is in progress. A white paper on Cognitive Radio has been produced [9] earlier and recently they reported a real time Cognitive Radio testbed for physical layer and link layer experiments [10]. Their motivation is to improve the spectrum utilization by opportunistic use of the spectrum, which is the same as the FCC's initiative. They treat the subject in a complete fashion: from physical layer issues to MAC layer issues and from analog frontend to computing platform supporting baseband processing. It is interesting to look into their study on spectrum sensing. From the system perspective, they study some basic considerations: the link budget of the sensing; the effect of noise on sensing; the cooperation of sensing nodes. From the signal processing perspective, a comparison study is made on different sensing techniques. Some hardware considerations related to sensing are also proposed see [11].

Spectrum pooling [6] is investigated by Jondral from the University of Karlsruhe. The basic idea is that a secondary user can dynamically access the licensed band by switching on and off OFDM subcarriers to avoid interference to the licensed user (primary user). However, the spectrum power leakage problem in the FFT based traditional OFDM systems could cause potential interference to the licensed system. They also observed this in [12], where two counter measures are proposed: spectrum shaping and switching off the carriers adjacent to the licensed user (primary user). There are other challenges for spectrum pooling like detection of spectrum access and synchronization, see [6].

The Cognitive Radio project [13] at Virginia Tech. does not specifically aim to improve spectrum utilization. This project is based on the observation that Cognitive Radio distinguishes itself by awareness and learning. In [14], a genetic algorithm based cognitive engine is proposed to learn its environment and respond with an optimal adaption. This approach to Cognitive Radio is more or less similar to the original concept of Mitola.

The European Union 6th framework End-to-End Reconfigurability (E2R) [15] project brings together major European players in the domain of reconfigurability, software defined radio and cognitive radio. The key objective of the E2R project is to devise, develop, trial the architectural design of reconfigurable devices and supporting system functions for users, application and service providers, operators, and regulators in the context of heterogeneous systems. Although the project does not specifically address Cognitive Radio, dynamic spectrum allocation and evolution from software defined radio to Cognitive Radio has been envisioned.

In parallel with the ongoing research projects around the world, international standardization organizations also have proposals to improve the spectrum utilization. An example is IEEE 802.22 [16], which is a new standard for a cognitive point-to-point (P2P) air interface for spectrum sharing with television bands. Television channels are very suitable for cognitive radio because they have a relatively unique spectrum signature that is easy for a cognitive radio to identify. The signals are also rigidly assigned to 6-MHz-wide channels with fixed center frequencies.

In the scope of our AAF project, we focus on the spectrum awareness and access. The objective is to demonstrate a Cognitive Radio system for emergency networks. Each radio node works in an ad-hoc based network, the node adopts the AAF protocol stack which is consistent with the five-layer protocol reference model (physical layer, data link layer, network layer, transport layer and application layer). The physical layer is the heart of our study where the free spectrum is discovered. Whenever the free spectrum is found, the AAF system will create an infrastructure using this spectrum.

2   Spectrum Sensing

In this section an architecture of a generic sensing system is described which consists of a spectral analysis system and a decision system. As an example, we present a sensing system that uses energy detection in the frequency range between $400-500$ MHz. For disaster relief networking this range is interesting because of propagation conditions, power requirements and antenna dimensions.

A brief analysis into the gains to be expected by collaborative scanning, a form of distributed detection, completes this section.

2.1   Sensing System Architecture

In Figure 1 a functional architecture of a sensing system is presented. Basically the system consists of two parts: a spectral analysis system and a decision system. An option for the spectral analysis system is to use energy detection (power detection). In this case the scanning system works on any signal, however, without using all available knowledge of signal properties of the primary user. Another option is to use feature establishment - one identifies well known (deterministic) signal features of the primary user's signal like carrier waveforms or pilot tones. As, for instance in broadcast situations, the primary user wants to be heard, it is expected that especially in bad SNR conditions feature establishment may outperform energy detection [17]. In this paper we focus on energy detection, as it is applicable to *any* primary user signal and possibly less energy consuming than feature detection.

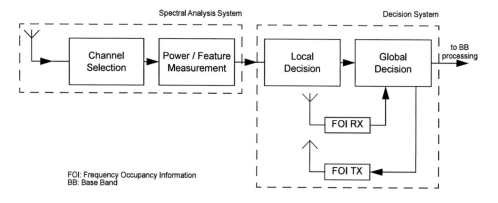

Figure 1: Functional Architecture of the sensing system.

A single scanning node needs to decide whether or not the band under consideration is empty or not, hence it takes a local decision. The quality of this decision (in terms of detection probability P_d and False Acceptance Rate (FAR)) may be hampered by two issues:

Multipath fading Signals that travel different signal paths may or may not add coherently at the receiver. As a result, small changes in location (relative to the wavelength λ) impact detection quality.

Shadowing An individual sensing node may be blocked by an obstacle and may therefore not be able to detect the primary user. Shadowing may occur over an area of 5 − 500 m, depending on the size of the object.

To overcome these problems, a collaborative scanning system can be used. In such a system Frequency Occupancy Information (FOI) is gathered by all individual scanning nodes and disseminated to the nodes participating in the secondary-user network using a special signalling channel for this purpose. The properties of this FOI dissemination channel (like its bandwidth, SNR, data rate, MAC protocol and delay), the independence of measurements and the number of nodes involved in the scheme all determine the gains in detection quality to be expected over locally-made-only decisions.

Leaving aside feature detection, the channel selection and power measurement part together form a spectrum analyser. Below we focus on alternative architectures for such a system.

2.2 Architectures of the Spectral Analysis System

We distinguish three architectures for the spectrum analysis part of a power-based sensing system: a scan-based architecture, a real-time FFT based architecture and an FFT-based architecture.

2.2.1 Scan-based Architecture

In Figure 2a, a scan-based architecture is presented. After a broad lowpass filter, a mixer driven by a local oscillator (LO) downconverts the signal to some intermediate frequency (IF). Subsequently the signal is anti-alias filtered, mixer products are removed and analog-to-digital conversion (ADC) is performed. In the digital realm a sample-and-hold type of filter is used and subsequently the energy in a particular frequency bin is computed. Instead of measuring frequency bins in parallel, they are measured sequentially, by changing the frequency of the LO and sweeping over the required frequency span (FS).

One critical design issue (especially for handheld nodes) is the power consumption of the AD converter. In order to compare the performance of AD converters a Figure of Merit (FoM) can be used that takes into account the converter's power dissipation P [W], its resolution (specified in Effective Number of Bits (ENOB)) and its sampling rate f_s [Hz], [18], [19]. This FoM is given by

$$\text{FoM} = \frac{P}{2^{\text{ENOB}} \cdot f_s} \; [\text{pJ} / \text{conversion-step}]. \tag{1}$$

in which a FoM of approximately 2 pJ / conversion-step is valid for today's ADCs. As dynamic ranges of as large as 80 dB can be observed for the 400 MHz-500 MHz frequency range ([20], [21, pp.77-84]) approximately $80/6 = 14$ bits ENOB are needed (we apply the well-known 6 dB/bit rule for ADCs), which severely limits the possible bandwidth that can be converted (say, to around $f_s/2 = 1 - 10\,\text{MHz}$, so power $P = 66 - 655\,\text{mW}$).

However, the main disadvantage of the sweeping approach is the huge loss of time resolution. First, it takes time for the filters to settle, which is inversely related to the resolution bandwidth (RBW). So the sweep frequency has to 'dwell' a certain amount of time in every frequency band. Also, the completion of one sweep takes an amount of time proportional to FS/RBW, so the sweep time is equal to

$$ST = k\frac{FS}{RBW^2} \tag{2}$$

in which k is a factor that depends on the filter type with typical values between 2 and 3, [22].

In Figure 2b the time-frequency diagram of the scan based approach is shown. In this figure the primary parameters involved in the scan-based method are indicated. The scanning starts one time t_s on frequency f_L. The scanning ends on t_e on frequency f_H. The complete scan time is ST, and the frequency span is FS. The Inter-bin frequency (IBF) is the distance between two adjacent frequency bins. The Inter-bin time (IBT) is the time it takes to complete a power measurement in a single time-frequency bin. The frequency bins are described by two parameters: The resolution bandwidth (RBW) and the dwell time (Td).

The gray bins are the time-frequency points that are 'seen' by the spectrum analyser. The bins that are not on the gray diagonal are just missed. Depending on the usage pattern of the primary user, this may or may not be a problem. In case of stable broadcast bands it seems less of a problem than in data-communication bands. In the example the number of bins is #points=8.

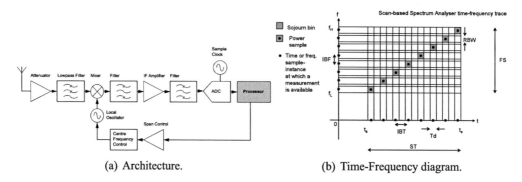

(a) Architecture.　　　　　(b) Time-Frequency diagram.

Figure 2: Scan-based approach.

2.2.2 Real-time FFT Based Architecture

Another approach is to calculate the spectrum with a discrete Fourier transform (DFT) from time domain measurements instead of measuring it directly.

Since we want to perform the Fourier transform digitally we first have to convert the RF signal to baseband using a quadrature mixer, apply an anti aliasing filter, remove mixing products and apply analog-to-digital conversion. A digital signal $u[n]$ has a discrete Fourier transform $U[k]$ defined as:

$$U[k] = \sum_{n=0}^{N-1} u[n]\, e^{-j\frac{2\pi}{N}kn} \quad \text{with } k = 0, \ldots, N-1 \tag{3}$$

The conversion from time domain to frequency domain can be done with the very efficient FFT (Fast Fourier Transform) algorithm which takes only about $N\log(N)$ multiplications instead of the N^2 multiplications for a general DFT. In Figure 3a, the architecture of the FFT based method is shown. The front end is tuned on a fixed frequency. The channel select filter is replaced by a much wider filter that passes the whole frequency span. The selection of the frequency bins is now done by the FFT which processes all channels in parallel.

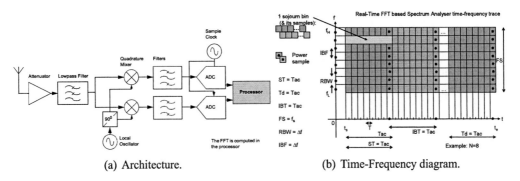

(a) Architecture.   (b) Time-Frequency diagram.

Figure 3: FFT-based approach.

The primary parameters involved in the FFT-based approach are given in 3b. If we compare this figure with Figure 2b we see some important differences. First we see that now all bins are covered instead of just one diagonal. So there are no points in the time-frequency plane that are missed. We also see that the measurement results for all frequency bins become available at the same time, after the scan time ST, which is equal to the acquisition time if we neglect the processing time ($Tac = NT$, with T the sampling time). The frequency span is equal to the sample frequency $FS = f_s = 1/T$.

The real-time FFT based method has drawbacks too. The major disadvantage of the FFT method is that two ADCs is needed, which cause even more power-related problems as was the case for the scan-based architecture. For instance, an FS of 100 MHz cannot be accommodated. However, in the UHF/VHF region bandwidths of $1 - 10$ MHz seem feasible.

2.2.3 Scanning-FFT Based Architecture

A scan-based architecture is more energy-efficient compared to a real-time FFT based one, but considered too slow for our purposes. The real-time FFT based architecture is fast but too power intensive for a large frequency span, so the natural way to proceed is to combine both methods. One chooses a feasible FFT size (determined by the power requirements of the ADC) with a frequency span that is (thus) a fraction of the total span required. We then can use the FFT multiple times and adjust the centre frequency of the front end each time so that the whole span is covered. In Figure 4 the resulting architecture and the time-frequency diagram of this configuration are given (in the example, F=2 tuning steps of the LO are needed).

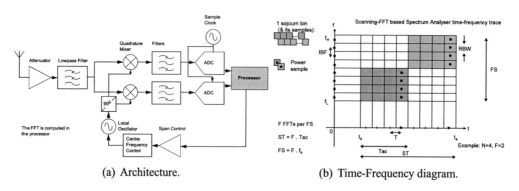

(a) Architecture.   (b) Time-Frequency diagram.

Figure 4: Scanning-FFT based approach.

To provide some quantitative insight we provide an example. Assume we require a total frequency span of $FS = 100\,\text{MHz}$ with a resolution of $RBW = 10\,\text{kHz}$, so 10000 bins. If we want to process this with a radix-2 FFT we need to sample and compute an 16384 points FFT every $Tac = 100\,\mu s$ (the first power of two larger than 10000). This makes up a substantial computational load.

If we use a scan-based architecture, according to equation (2) with k=2, it will take the sweep $2\frac{100\cdot10^6}{(10\cdot10^3)^2} = 2\,\text{s}$ to complete, which may be too long if we want to detect within a MAC frame duration (in the order of milliseconds).

If we use a scanning-FFT based architecture, we can partition the total span in 20 blocks of 5 MHz, which may be processed by calculating 512 points every $100\,\mu s$. And since the bandwidth is reduced by a factor 20, the sampling rate and thus the power consumption is also reduced by the same factor (see (1)).

Another benefit of a scanning-FFT based architecture is that the LO can interplay independently with a hopping baseband processing system, as was suggested in [23].

2.3   Specification Parameters for the Spectral Analysis System

The spectral analysis system can be described by the specification parameters of a spectrum analyzer as given in table 2. The meaning of the parameters is illustrated in figures 2b, 3b and 4b. Note that the parameters are not independent of one another and, moreover, the dependency depends on the chosen scan-system architecture (see [22], [24] and [25]).

As an example, consider a frequency range of interest from 400 MHz to 500 MHz. In order to see what current-day capabilities of energy-based sensing systems are, we investigate the capabilities of existing commercially available spectrum analysers (Agilent 4407B, Rohde & Schwarz FSP3 and Tektronics FSA3308A). Below, we present some parameter-*values* of these systems.

- The required frequency resolution of the scanning system can be taken from the

Table 2: Specification Parameters for the Spectral Analysis System ([22], [24], [25]).

Aspect	Parameter	Unit	Variable	Remarks
Frequency	Frequency Span	[Hz]	FS	
	Resolution Bandwidth	[Hz]	RBW	
	Inter-bin Frequency	[Hz]	IBF	
Time	Sweep Time	[s]	ST	
	Dwell Time	[s]	Td	A.k.a. Sojourn time, S&H integration time, Time per bin
	Inter-bin Time	[s]	IBT	
Power	Dynamic Range	[dB]	ΔdB	
	Sensitivity	[dBm]	DANL	Displayed Average Noise Level
Memory	#points	-	Number of bins	

Dutch National Frequency Plan [26] and from the Dutch frequency register [27]. In the range we studied (400 MHz - 500 MHz) a resolution bandwidth RBW=10 kHz seems sufficient, as the nominal channel spacing ranges from 12.5 kHz to 25 kHz (below 470 MHz) and from 200 kHz to 8 MHz above 470 MHz (TV bands).

- For a frequency span FS=100 MHz and a resolution bandwidth RBW=1 kHz the shortest sweep-time ST equals 8 s (using a Rohde & Schwarz FSP3 with #points=8000 and "FFT-filters" mode). This seems too long for MAC-frame durations (e.g. 2 ms for HiperLAN/2), but may be sufficient for course-grain measurements (e.g. with RBW's in the range of MHz's the sweep times (STs) become smaller).

- The studied spectral analysers do not provide high time-resolution in combination with high frequency resolution spectral measurement (the maximum real-time FFT bandwidth was 15 MHz (Tektronics FSA3308A), so FS=15 MHz).

- All studied spectrum analysers have a dynamic range of around 80 dB.

- The required time resolution of the scanning system depends on the utilisation pattern of the primary users.

As can be seen, a good tradeoff has to be found between time resolution, frequency resolution and frequency span. In case high frequency resolution is needed and the scanning needs to take place on the time scale of a MAC frame (milliseconds) a real-time FFT based architecture can be used, however with a limitation on the frequency span. A change in carrier frequency of the scanning system seems possible per MAC frame, so that the preferred architecture is a scanning-FFT based one.

2.4   Collaborative Scanning: Distributed Detection

In this secction we propose a decision system for the detection of a primary user signal. Suppose the decision system (see Figure 1) takes a local hard-decision (a yes/no one) on an energy-based measure. So, for the k-th frequency bin

$$E_k = |U[k]|^2 \qquad (4)$$

is computed with $U[k]$ as in (3). Assume $U[k] = X[k] + W[k]$ in which $\{X[k]\}$ is the (complex-valued) primary user signal in the k-th frequency bin (which can be anything, with unknown distribution) and $\{W[k]\}$ is assumed to be (complex-valued) white Gaussian noise, independent of the primary user's signal. The expected value of the signal is $E[|U[k]|^2] = E[|X[k]|^2] + E[|W[k]|^2]$. Now, we formulate two hypotheses H_i:

H_0 :   *no* primary user signal is present, so $|U[k]|^2 = |W[k]|^2$, and

H_1 :   a primary user signal *is* present.

Subsequently we define a decision rule D:

$$D_k = 0 \text{ in case } E_k < \theta \text{ and } D_k = 1 \text{ in case } E_k > \theta$$

in which D_k is the decision made by the local decision system for the k-th frequency bin and θ is the detection threshold.

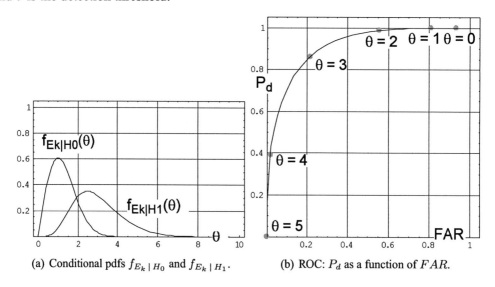

(a) Conditional pdfs $f_{E_k \mid H_0}$ and $f_{E_k \mid H_1}$.

(b) ROC: P_d as a function of FAR.

Figure 5: Detection quality of a single node.

The quality of the detection can be expressed in two conditional probabilities, the detection probability P_d and the false alarm rate FAR. The probability of detecting a primary-user signal is given by $P_d \triangleq Pr[D_k = 1 \mid H_1]$, the probability of detecting noise instead of a primary-user signal is $FAR \triangleq Pr[D_k = 1 \mid H_0]$. In case the probability density functions (pdfs) of E_k under both hypotheses are available, so $f_{E_k \mid H_1}(x)$ and $f_{E_k \mid H_0}(x)$ are known, the detection probability and false alarm rate can be computed as

$$P_d = \int_\theta^\infty f_{E_k \mid H_1}(x)\, dx$$

$$FAR = \int_\theta^\infty f_{E_k \mid H_0}(x)\, dx.$$

(5)

A graph of P_d versus FAR is called a Receiver Operating Curve (ROC), see [28]. An example for the pdfs in Figure 5a is given in Figure 5b. Observe that it is in fact the choice of the threshold value θ that determines a position on the curve in the graph. What a good choice of P_d and FAR is, is determined by *other* criteria like the cost of causing interference to a primary user (rudeness), or the cost of finding too little unoccupied spectrum (weakness).

Can the detection quality for the k-th frequency bin be improved by combining the hard decisions of N nodes for this bin?

We assume that a particular sensing node receives the decisions of the $N-1$ other nodes through the FOI channel. The question is now how to combine these N local decisions into a global one (aiming at better quality). A possible fusion rule is a rule which decides H_1 if *any* of the N received local decisions is 1 (the 1-out-of-N rule or OR-rule). For this fusion rule the detection probability $P_d^{(N)}$ is [29]:

$$P_d^{(N)} = 1 - (1 - P_d)^N$$

(6)

and the $FAR^{(N)}$ is

$$FAR^{(N)} = 1 - (1 - FAR)^N.$$

(7)

In Figure 6 the effects of this rule and the number of collaborative nodes are illustrated. By increasing N it can be seen that the ROC curves creep into the upper left corner, signifying improved detection quality.

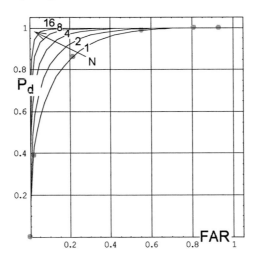

Figure 6: ROC for collaborative scanning (N nodes).

Observe that in this analysis we assumed that all nodes independently "see the same thing (primary user's frequency occupancy)". Apart from the fusion rule shown here, other rules may be applied that result in a different detection quality for the collaborating nodes. In case nodes are in disagreement (e.g. because of fading or shading they do *not* "see the same thing") another analysis is necessary and another fusion rule may outperform the one sketched here.

The approach assumes a more or less errorless instantaneous availability of FOI information for the global decision function (Figure 1). As such the values found by this method of analysis have to be interpreted as a *benchmark* for practical collaborative sensing schemes in which (imperfect) FOI dissemination channels are used.

3   OFDM Based Cognitive Radio Baseband System

In this section we present adaptive OFDM based schemes for Cognitive Radio. We will first discuss adaptive bit loading and adaptive power loading. Second, we will discuss the basic OFDM baseband processing functions in an OFDM receiver.

3.1   Adaptive Bit Loading and Adaptive Power Loading

The basic idea of Orthogonal Frequency Division Multiplexing (OFDM) is to transmit high data rate information by dividing the data into several parallel bit streams where these bit streams are modulated by orthogonal frequencies. The benefit of OFDM is that the high data rate of the whole system is transformed into relatively low data rate streams on each subcarrier which is more robust to inter symbol interference (ISI) caused by multipath delay spread. In the hardware design, it is expected that an OFDM radio is also easy to integrate with spectrum sensing because both of them use FFT cores. The system robustness and hardware resource sharing make OFDM a good choice for Cognitive Radio baseband systems. More importantly, an OFDM system can optimally approach the Shannon capacity in the segmented spectrum by adaptive resource allocation on each subcarrier, which includes adaptive bit loading and adaptive power loading [30]. In an OFDM based Cognitive Radio system, information bits are loaded as different modulation types onto each available subcarrier depending on the subcarrier's signal to noise ratio while the subcarriers currently not available to Cognitive Radio are switched off. Two optimization methods for the adaptive resource allocation can be used for Cognitive Radio: using the power constraint or using the data-rate constraint.

We could maximize the data rate of the system under a certain power constraint. It is formulated as follows:

$$Max \ R = \sum_{k=1}^{K} \frac{F_k}{K} \log_2(1 + \frac{h_k^2 p_k}{N_0 \frac{B}{K}})$$

$$Subject \ to: \sum_{k=1}^{K} p_k \leq P_{total}$$

$F_k \in \{0, 1\}$ for all k

$p_k = 0$ for all k which satisfies $F_k = 0$ \hfill (8)

where R is the data rate; K is the number of the subcarriers. N_0 is the noise power density, B is the band of interest for Cognitive Radio, h_k is the subcarrier gain and p_k is the power allocated to the corresponding subcarrier. F_k is the factor indicating the availability of subcarrier k to Cognitive Radio, where $F_k = 1$ means the kth carrier can be used by Cognitive Radio.

The system power minimization can also be applied under the constraint of a constant data rate. We formulate it as follows:

$$Min \ \sum_{k=1}^{K} p_k = P_{total}$$

$$Subject \ to: R = \sum_{k=1}^{K} \frac{F_k}{K} \log_2(1 + \frac{h_k^2 p_k}{N_0 \frac{B}{K}})$$

$F_k \in \{0, 1\}$ for all k

$p_k = 0$ for all k which satisfies $F_k = 0$ \hfill (9)

A simulation result of adaptive bit loading is shown in Figure 7.

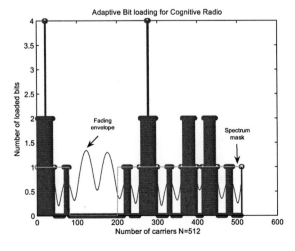

Figure 7: Adaptive bit loading for Cognitive Radio

Table 3: Properties of the different OFDM standards

	Hiper LAN/2	DAB I	DAB II	DAB III	DAB IV	DRM A	DRM B
Operating frequency[Hz]	5G-6G	174M-240M	1.4G	<3G	1.4G	<30M	<30M
Bandwidth[MHz]	20	1.54	1.54	1.54	1.54	0.012	0.012
Used carrier number	52	1536	384	192	768	226	206
Symbol time [μs]	4	1246	312	156	623	26667	26667
Frame time [ms]	2	96	24	24	48	400	400
FFT size	64	2048	512	256	1024	288	256

Table 4: DRM transmission modes

parameter	mode A	mode B	mode C	mode D
$T_u[ms]$	24	21 1/3	14 2/3	9 1/3
$T_g[ms]$	2 2/3	5 1/3	5 1/3	7 1/3
FFT size	288	256	176	112
used carriers	226	206	138	88

In the simulation, we assume a constant data rate of 400 bits for 512 subcarriers in a frequency selective channel. A binary spectrum mask is given by the sensing system to indicate the spectrum availability, a "1" means the subcarrier is available to Cognitive Radio. The optimization method in eq. (9) is applied to achieve constant data rate with power optimization.

3.2   OFDM Baseband Processing

OFDM is widely used in current standards such as HiperLAN/2 [31], DAB [32] and DRM [33]. Due to various system requirements and different radio propagations on different frequency bands, OFDM standards vary a lot in the number of carriers and the transmission bandwidth. Table 3 summarizes the OFDM properties for different standards. Within one standard, multiple modes can be used to cope with various transmission channels by adopting different parameter sets. We take DRM as an example (see Table 4). Mode A is applied for transmissions over Gaussian channels or channels with minor fading which are typical for ground-wave transmissions (medium wave, long wave). The other modes can cope with more time and frequency selective channels which are typical for sky-wave transmissions (short wave and medium wave at night). Mode B and C have the same degree of robustness against delay spread, however mode C has better robustness against Doppler spread because of a shorter useful time T_u. Mode D has the highest robustness against severe delay and Doppler spread among all DRM modes at the cost of the lowest data rate. As we mentioned, Cognitive Radio can operate in different frequency bands, provide multimedia services with various QoS and cope with different channel conditions, therefore Cognitive Radio is able to adapt to different OFDM systems.

Although there are a lot of differences in various OFDM systems, the basic baseband processing is rather similar. This makes OFDM based Cognitive Radio feasible. An OFDM baseband receiver generally consists of the following basic tasks (Figure 8):

- **Packet/frame detection and synchronization** is used to determine the starting point of an OFDM frame. This is usually achieved by correlating the received signal with known preambles. We also refer to this frame synchronization as coarse synchronization in contrast with the fine synchronization of the OFDM symbols.

- **Frequency offset estimation and correction** are done to remove the frequency offset which destroys the frequency orthogonality of subcarriers. Frequency offset estimation is done on the frame basis in the preamble section (see Figure 8) before data symbols. A frequency correction coefficient determined by the frequency offset estimation is multiplied by each OFDM data symbol.

- **Channel estimation and equalization** are used to correct the frequency selective fading. Due to the robustness of OFDM to frequency selective fading, less complex frequency domain equalization techniques can be used.

- **Guard time removal** is done after the fine synchronization of OFDM symbols.

- **FFT** is the basic component of all OFDM systems which transforms the signal into the frequency domain.

- **Phase offset tracking and correction** correct the frequency offset residual error by using the pilots in each OFDM symbol.

- **De-map** transforms the complex numbers to a bitstream according to their modulation types. Different modulation types can be loaded on each subcarrier by applying adaptive bit-loading as we have mentioned.

- **De-interleaving** is the opposite task of data interleaving. The idea of data interleaving is that the original data is re-arranged at a transmitter in order to reduce burst errors on a receiver. De-interleaving is the inverse operation of interleaving.

- **Channel decoding** exists in all communication systems not only in OFDM systems. However, OFDM is almost always used in conjunction with channel coding to create coded OFDM (COFDM). A Viterbi code is commonly used, but other channel codes such as turbo code or a LDPC code can also be applied. The code rate can be adaptive to provide different degrees of error protection.

For simplicity we only mentioned the tasks performed by the receiver. The transmitter basically does the inverse operations which are less complex. The aforementioned functional blocks can be found in all OFDM systems, but different standards may use different algorithms for each functional block. This means the communication system can select an algorithm to perform each function depending on the requirements of the system. For example, different channel encoders/decoders (codecs) can be applied to achieve different QoS requirements. We refer to this adaptivity as *algorithm-selection level* adaptivity [34]. For a specific algorithm, there are also opportunities for adaptivity by changing parameters

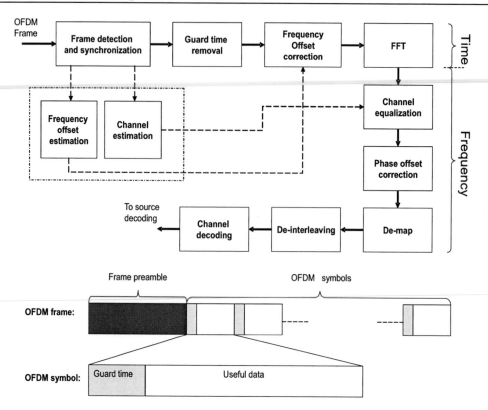

Figure 8: Basic OFDM processing tasks

of the algorithm. For example the size of FFT and the code rate of Viterbi codec can be tuned by different standards or modes. We refer to this adaptivity as *algorithm-parameter level* adaptivity. The next question is how to achieve the adaptivity for different OFDM systems. A fully reconfigurable hardware platform will be an ideal solution.

4 Heterogeneous Reconfigurable Platform

As already foreseen by Mitola [1], a Cognitive Radio is the final point of software-defined radio platform evolution: a fully reconfigurable radio that changes its communication functions depending on network and/or user demands. His definition on reconfigurability is very broad and we therefore focus on a reconfigurable hardware platform which supports Cognitive Radio. In this section, we present a heterogeneous reconfigurable hardware platform for Cognitive Radio.

4.1 A Heterogeneous Reconfigurable System-on-Chip Architecture

With the evolution of semiconductor technology, more and more transistors can be integrated on a single chip which makes it possible to build large systems, on a chip level rather than on a board level. This approach is called System-on-Chip (SoC). The reconfigurable platform we propose for Cognitive Radio is a heterogeneous reconfigurable SoC architec-

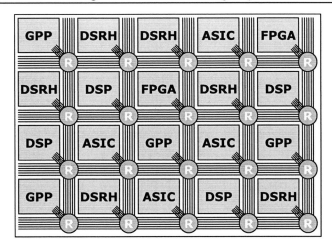

Figure 9: An example of a heterogeneous System on Chip (SoC). DSRH = Domain Specific Reconfigurable Hardware

ture shown in Figure 9.

This SoC is a heterogeneous tiled architecture, where tiles can be various processing elements including General Purpose Processors (GPPs), Field Programmable Gate Arrays (FPGAs), Application Specific Integrated Circuits (ASICs) and Domain Specific Reconfigurable Hardware (DSRH) modules. The tiles in the SoC are interconnected by a Network-on- Chip (NoC). Both the SoC and NoC can be dynamically reconfigurable, which means that the programs (running on the reconfigurable processing elements) as well as the communication links between the processing elements are configured at run-time. Different processing elements are used for different purposes. The general purpose processors are fully programmable to perform different computational tasks, but they are not energy-efficient. The dedicated ASICs are optimized for power and cost. However, they can not be reconfigured to adapt to new applications. FPGAs which are reconfigurable by nature, are good at performing bit-level operations but not that efficient for word level DSP operations. The Domain Specific Reconfigurable Hardware (DSRH) is a relatively new type of processing element, where the configurable hardware is tailored towards a specific application domain. The Montium [35] tile processor (see Figure 10) developed at the University of Twente, and recently commercialised by Recore Systems, is an example of DSRH. It targets the digital signal processing (DSP) algorithm domain, which is the heart of the wireless baseband processing. In our previous work [36] [37] [38], several DSP algorithms used in wireless communication have been mapped onto the Montium architecture. The implementation results show that the Montium architecture is flexible enough to adapt to different algorithms with good energy-efficiency. In a broader sense, working intelligently in an energy-efficient way is an important feature of Cognitive Radio. For Cognitive Radio devices working in the emergency network, energy-efficiency is really a crucial issue because the battery life of radio devices can be a limitation for successful operations. Therefore the reconfigurable platform we propose not only targets flexibility but also energy-efficiency. In the next section, we will introduce the key element of this platform, the Montium tile processor.

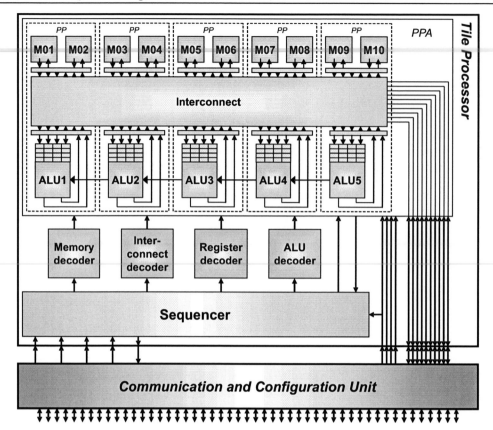

Figure 10: An example of DSRH: a Montium processor

4.2   The Montium Tile Processor

The Montium is an example of DSRH which targets the 16-bit digital signal processing (DSP) algorithm domain. At first glance the Montium architecture bears a resemblance to a Very Long Instruction Word (VLIW) processor. However, the control structure of the Montium is very different. For (energy-) efficiency it is imperative to minimize the control overhead. This can be accomplished by statically scheduling instructions and using instruction decoders. The lower part of Figure 10 shows the Communication and Configuration Unit (CCU) and the upper part shows the reconfigurable Tile Processor (TP). The CCU implements the interface for off-tile communication. The definition of the off-tile interface depends on the NoC technology that is used in the SoC. The CCU enables the Montium to run in 'streaming' as well as in 'block' mode. In 'streaming' mode the CCU and the Montium run in parallel (communication and computation overlap in time). In 'block' mode the CCU first reads a block of data, then starts the Montium, and finally after completion of the Montium, the CCU sends the results to the next tile. The TP is the computing part that can be configured to implement a particular algorithm.

The Montium architecture has been refined to perform common DSP algorithms used in wireless communication. Results in [36] showed the Montium's adaptivity to several DSP algorithms including correlation, Finite Impulse Response (FIR) filtering and FFT.

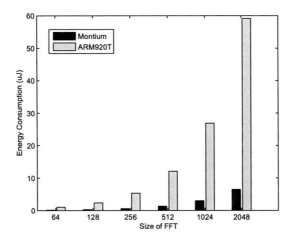

Figure 11: Energy consumption (in μJ) comparison between the Montium and ARM

Later in [37] an adaptive Viterbi algorithm, considered as the most energy consuming part in wireless baseband processing, was mapped onto the Montium. Recently an irregular non-power-of-two FFT has been mapped [38]. Research is still going on to investigate the Montium's adaptivity to new algorithms.

Besides it's flexibility, the Montium is very energy-efficient. By statically scheduling instructions at compile time, the overhead of both communication and control is reduced. A good example, presented in [35], is that a FIR filtering algorithm implemented on the Montium does not change the instructions in 99% of the time. Therefore, the instruction decoding does not result in excessive switching of control signals. To give a better idea how energy-efficient the Montium is, we show the implementation results of the FFT. This algorithm is one of the most computationally intensive parts, therefore the most energy consuming in OFDM baseband processing. It is common practice to implement these algorithms on dedicated ASICs, however this dedicated hardware can not be reconfigured because, for example, the size of the implemented FFT is fixed. The OFDM based Cognitive Radio we conceive has to be flexible. Therefore a dedicated ASIC is not a solution.

In the OFDM based Cognitive Radio system, an FFT is used by both OFDM baseband processing and spectrum sensing. The size of the FFT has to be adaptive to various OFDM systems. Moreover, spectrum sensing may apply coarse sensing or fine sensing with variable sized FFTs. Let us consider the most common radix-2 FFT. As the size of FFT increases, the computational complexity of the FFT increases on a logarithmic scale because the order of the complexity is $\frac{N}{2}\log_2 N$. So, FFT processing especially for large size FFTs have tough demands for the processor in terms of timing and energy consumption. A general purpose embedded solution like the ARM (a reduced instruction set computer (RISC) processor architecture) is inefficient for these type of applications. According to [35], the execution part of an FFT butterfly takes 21 clock cycles on an ARM920T running at 250MHz while it takes only 1 clock cycle on the Montium running only at 100MHz. Put in a simple way, a Montium is about ten times faster than an ARM. The energy consumption of the Montium is significantly lower than of the ARM, as illustrated in Figure 11.

For fair comparison, the ARM implementation is optimized and both implementations are implemented in 0.13 μm technology. If a 1024 points FFT is being executed for spectrum sensing while another 1024 points FFT is required for OFDM baseband processing at the same time, the energy saving of the Montium implementation will be significant compared with the ARM implementation as shown in Figure 11.

4.3   The Design Methodology

To implement Cognitive Radio on a heterogenous reconfigurable platform, we propose a design methodology which has two new features: 1) Transaction level modelling of an application into a parallel task graph, 2) Run-time spatial mapping of tasks onto heterogenous processing tiles.

First, a high level application such as the baseband processing part of a Cognitive Radio, is partitioned into tasks. The applications are typically streaming applications. The tasks within the applications communicate with each other via channels which are accessed by tasks through interfaces. Communications between tasks are data transactions including synchronization and data transfer. In a transaction level model, the details of communication among computation components are separated from the details of the computation components. For example an OFDM receiver can be modelled by the computation components (or tasks) we have mentioned in the previous chapter while communications are modelled at transaction level. In our implementation model, we use the Task Transaction Level (TTL) approach [39]. Here we give a pseudo code example of the FFT task implementation.

```
process Task_FFT {
  initialization;
  while (true)
  {\\ declaration of local variables
    complex Y[64],Z[64];
    for(i=0;i<64; i++)
    \\ read input samples
    InRead(&(Task_FFT->inport), &Y[i]);
    \\ perform the FFT
    call function Z=FFT(Y);
    for(i=0; i<64; i++)
    \\ write output samples
    OutWrite(&(Task_FFT->outport), &Z[i]);
  }
}
```

In principle, the FFT size is application dependent but in the example we assume the FFT size is 64. While the condition is true, 64 samples are read into the local buffer from the channel which is connected to the input port. When 64 samples are ready, data processing is executed by calling the FFT function. This FFT function can be implemented on various processing tiles like ARM, FPGA or Montium. Results are written to the channel connected to the output port. Both synchronization and data transfer are done by simple read and write function calls. From the example we can see that the computation components can be plugged-in and replaced as functions, which allows application adaptivity within the same TTL framework. So, it is advantageous to use the transaction level model to speed up simulation and allow exploring and validating design alternatives at a high level of abstraction.

For each task of an application, one or more task implementations have to be provided. A task implementation is the implementation of a task on a particular tile, e.g. object code for an ARM or a DSP or configuration data for an FPGA or a Montium. A task implementation has several characteristics, e.g. the amount of energy it takes to execute the task on a particular tile of the architecture, the delay or tile utilization. This implementation library (including the characteristics) is composed at design-time.

At run-time, a spatial mapping function module selects the tile implementation for each task of an application. The objective of mapping is to find a mapping solution which optimizes a certain design metric such as the energy consumption given other system constraints (e.g. timing, tile availability). Such a mapping problem is known to be NP-hard and in practice, the corresponding optimization algorithms may run for hours. Therefore, heuristics have to be used to find a solution with a reasonable quality within an acceptable time. In [40], a minimal weight algorithm is proposed to map a digital audio broadcast receiver onto a tiled SoC in a reasonable time, where the optimization metric is to minimize the energy consumption. Run-time mapping offers a number of advantages over design-time mapping, especially for a Cognitive Radio which dynamically changes its behavior.

- First, run-time mapping can adapt to the available resources. The available resources may vary over time due to multiple applications running simultaneously.

- Second, run-time mapping can adapt to dynamically changing applications. A task within the application may change its computational behavior at run-time, for example, the size of the FFT or a new task can be added to the application.

5 Conclusion

In this chapter, we give an introduction into Cognitive Radio in the context of emergency networks. Our focus on Cognitive Radio is to search for under-utilised spectrum and to (rapidly) adapt transmission settings accordingly to meet various system requirements of emergency networks. Therefore, the research on Cognitive Radio concentrates on frequency scanning and adaptive baseband processing. Three architectures for single-node energy-based scanning have been identified: scan-based, real-time FFT based and scanning-FFT based. The scanning-FFT based approach is shown to be most promising. By introducing collaboration between sensing nodes, an increase in detection quality can be expected.

The robustness to multipath delay spread and the hardware resource sharing with scanning FFT-based spectrum sensing, make OFDM a good choice for the Cognitive Radio baseband system. Moreover, OFDM systems that can nullify individual carriers poses interesting opportunities for Cognitive Radio. A highly adaptive OFDM system is foreseen since it has to operate in different frequency bands, it has to deal with various effects of wireless channels like fading and shadowing and it has to support various multimedia services. In order to support a flexible Cognitive Radio, a heterogenous reconfigurable SoC platform is required, similar to a Software Defined Radio. The key element in the platform, the coarse-grain reconfigurable tile processor (the Montium), enables the reconfigurability in combination with the energy-efficiency. A design methodology is needed to map applications onto a heterogeneous platform which has two new features: transaction level modelling of applications and run-time spatial mapping.

References

[1] J. Mitola III. *Cognitive Radio: An Integrated Agent Architecture for Software Defined Radio.* PhD thesis, Royal Institute of Technology, Sweden, May. 2000.

[2] S. Haykin. "Cognitive Radio: Brain-empowered Wireless Communication". *IEEE J. Select. Areas Commun.*, vol. 23, no.2:pp 201–220, Feb. 2005.

[3] http://www.freeband.nl/project.cfm?id=488.

[4] M. A. McHenry. "NSF spectrum occupancy meassurements project summary". http://www.sharespectrum.com.

[5] Tewfik L. Doumi. "Spectrum Considerations for Public Safety in the United States". *IEEE Commun. Mag.*, Jan. 2006.

[6] T.A. Weiss and F.K. Jondral. "Spectrum Pooling: An Innovative Strategy for the Enhancement of Spectrum Efficiency". *IEEE Commun. Mag.*, vol. 24, no.3:pp S8–S14, Mar. 2004.

[7] J. Mitola. "Cognitive Radio: Making software radios more personal". *IEEE Pers. Commun.*, Aug. 1999.

[8] Federal Communication Commision. "Spectrum Policy Task Force". Tech. Report., Nov. 2002.

[9] Robert W. Brodersen, Adam Wolisz, Danijela Cabric, Shridhar Mubaraq Mishra, Daniel Willkomm. "A Cognitive Radio Approach for Usage of Virtual Unlicensed Spectrum". White Paper, July 2004. `http://bwrc.eecs.berkeley.edu/Research/MCMA/CR_White_paper_final1.pdf`.

[10] Shridhar Mubaraq Mishra et al. "A real time Cognitive Radio testded for physical and link layer experiments". In *Proc. of the IEEE Symposium on New Frontiers in Dynamic Spectrum Access*, pages 562–567, Nov. 2005.

[11] http://bwrc.eecs.berkeley.edu/Research/Cognitive.

[12] T. Weiss, A. Krohn, J. Hillenbrand, and F. Jondral. "Mutual Interference in OFDM-based Spectrum Pooling Systems". In *Proc of VTC*, Milan, Italy, May 2004.

[13] http://www.cwt.vt.edu/research/cognitiveradio/index.html.

[14] Christian James Rieser. *Biologically Inspired Cognitive Radio Engine Model Utilizing Distributed Genetic Algorithms for Secure and Robust Wireless Communications and Networking.* PhD thesis, Virginia Tech.

[15] http://e2r.motlabs.com.

[16] http://www.ieee802.org/22.

[17] D. Cabric, S.M. Mishra, and R.W. Brodersen. "Implementation Issues in Spectrum Sensing for Cognitive Radios". 38th Asilomar conference on signals, systems and computers, pages 772–776, 2004.

[18] M. Pelgrom. "Analog-to-digital Conversion", March 2003. Course Notes University of Twente, course 121048.

[19] K.H. Lundberg. "High-Speed Analog-to-Digital Converter Survey", October 2002. Accessible via http://web.mit.edu/klund/www/papers/.

[20] F.H. Sanders and V.S. Lawrence. "Broadband Spectrum Survey at Denver, Colorado". NTIA Report TR-95-321, NTIA, September 1995. http://www.its.bldrdoc.gov/pub/surv_dnv/.

[21] M. McHenry and D. McCloskey. "New York City Spectrum Occupancy measurements, September 2004". Report, Shared Spectrum Company, December 2004.

[22] Agilent. "Spectrum Analyser Basics". Application Note 150, Agilent, 2004.

[23] F.W. Hoeksema, M. Heskamp, R. Schiphorst, and C.H. Slump. "A Node Architecture for Disaster Relief Networking". 1st IEEE International Symposium on new Frontiers in Dynamic Spectrum Access Networks(DySPAN'05), pages 577–584, November 2005.

[24] Agilent. "Swept and FFT Analysis". Performance spectrum analyzer series - application note, Agilent, 2004.

[25] Tektronics. "Fundamentals of Real-Time Spectrum Analysis". Technical report, *Tektronics*, 2004.

[26] Radiocommunications Agency The Netherlands) Dutch Ministry of Economic Affairs (Agentschap Telecom. "Nationaal Frequentieplan 2005 (National Frequency Plan 2005)". http://www.ez.nl/content.jsp?objectid=24642, February 2005.

[27] Radiocommunications Agency The Netherlands) Dutch Ministry of Economic Affairs (Agentschap Telecom. "Nationaal Frequentieregister (National Frequency Register)". http://www.agentschap-telecom.nl/nfr/, March 2005.

[28] H.L. van Trees. *Detection, Estimation, and Modulation Theory*, volume I: Detection, estimation, and Linear Modulation Theory. John Wiley & Sons, 1968.

[29] P.K. Varshney. *Distributed Detection and Data Fusion.* Springer-Verlag, 1997.

[30] I.Budiarjo, H.Nikookar, and L.P. Ligthart. "Overview of Adaptive OFDM in the Context of Cognitive Radio". In *12'th IEEE Symposium on Communications and Vehicular Technology in the BENELUX*, Enschede, the Netherlands, Nov. 2005.

[31] ETSI. "Broadband Radio Access Networks (BRAN); HIPERLAN Type 2; Physical (PHY) layer". Technical Specification ETSI TS 101 475 V1.2.2 (2001-02), 2001.

[32] ETSI. "Radio broadcasting systems: Digital Audio Broadcasting to mobile, portable and fixed receivers". Technical Specification EN 300 401.

[33] ETSI. "Digital Radio Mondiale (DRM) System Specification". Technical Specification ETSI TS 101 980 V1.1.1 (2001-09), 2001.

[34] Gerard K. Rauwerda, Jordy Potman, Fokke W. Hoeksema, and Gerard J.M. Smit. "Adaptation in the physical layer using heterogeneous reconfigurable hardware". In *Adaptation Techniques in Wireless Multimedia Networks*, 2006. ISBN 1-59454-883-8.

[35] Paul M. Heysters. *Coarse-Grained Reconfigurable Processors; Flexibility meets Efficiency*. PhD thesis, University of Twente, September 2004. ISBN 90-365-2076-2.

[36] Paul M. Heysters and Gerard J.M. Smit. "Mapping of DSP Algorithms on the Montium Architecture". In *Proceedings of Reconfigurable Achitecture Workshop*, 2003.

[37] Gerard K. Rauwerda, Gerard J.M. Smit, and Werner Brugger. "Implementing an Adaptive Viterbi Algorithm in Coarse-Grained Reconfigurable Hardware". In *Proceedings of the International Conference on Engineering of Reconfigurable Systems and Algorithms*, 2005.

[38] Arnaud Rivaton, Jérôme Quevremont, Qiwei Zhang, Pascal T. Wolkotte, and Gerard J.M. Smit. "Implementing Non Power-of-Two FFTs on Coarse-Grain Reconfigurable Architectures". In *Proceedings of the International Symposium on System-on-Chip (SoC 2005)*, 2005.

[39] Pieter van der Wolf et al. "Design and Programming of Embedded Multiprocessors: An Interface-Centric Approach". In *Proceedings of the CODES+ISSS*, 2004.

[40] L.T. Smit, J.L. Hurink, and G.J.M. Smit. "Run-time mapping of applications to a heterogeneous SoC". In *IEEE 2005 International Symposium on System-on-Chip Proceedings*, Tampere, Finland, 2005.

In: Mobile Multimedia: Communication Engineering ...   ISBN 1-60021-207-7
Editor: I.K. Ibrahim and D. Taniar pp. 101-116   © 2006 Nova Science Publishers, Inc.

Chapter 6

SECURING MOBILE COMMUICATION: RADIUS IN A WINDOWS ENVIRONMENT

Edgar R. Weippl* and Jamil Wahbeh
Security Research

Abstract

Security in Wireless-LAN systems is more than ever in the focus of the public. The focus of this chapter is on setting up a Wireless-LAN, which complies with the highest security requirements regarding authentication, authorization and data encryption. A successful authentication process is mandatory for the client to access the LAN physically after the authorization. The preceding authentication will be performed mutually at both communication partners. Furthermore, different Wireless-LAN user groups should be distinguished. They should be able to consume different authentication methods; every user group will use different authentication- and encryption-methods.

1   Introduction

Security in Wireless-LAN systems is more than ever in the focus of several working- and research groups (Wi-Fi, IEEE 802.11, ...) and is also important for Chief Security Officers. While there are a lot of theoretical possibilities to secure a Wireless-LAN, the practical implementation is lagging behind. The lack of user skills and the enormous amount of standards and working drafts, which are trying to solve the concept of security in Wireless-LANs, are the main reasons for that circumstance.

The focus of this chapter is on setting up a Wireless-LAN, which complies with the highest security requirements regarding authentication, authorization and data encryption. A successful authentication process is mandatory for the client to access the LAN physically after the authorization. The preceding authentication will be performed mutually at both communication partners. Furthermore, different Wireless-LAN user groups should be distinguished. They should be able to consume different authentication methods; every user group will use different authentication- and encryption-methods.

---

*E-mail address: weippl@securityresearch.at

They requirements of a WLAN, which meets the highest security criteria, can be defined as followed:

- Secure authentication

- Secure encryption

- 'Rogue AP' protection

- Different security layers with their corresponding authentication and authorization methods

The authentication should be processed by a RADIUS-server, which is installed on a Windows 2003 server platform. RADIUS (over EAP) will be used as the communication protocol between the access point and the RADIUS server. That means that EAP communication packets will be embedded in the RADIUS-protocol. EAP-TLS, which is able to guarantee a mutual authentication of both communication partners, will be used in addition to 802.1X between the client and the access point. Nowadays, EAP-TLS is the most secure communication protocol. TLS supports mutual authentication and is the successor of the popular SSL protocol. The RADIUS-server organizes the different roles in the network and assigns different rights to the users. To guarantee at the authentication a high degree of security, digital certificates will be used by both communication partners. So it is possible for both sides to validate the originality of the corresponding partner (this process is supported by the TLS protocol).

IEEE802 is the base of IEEE802.11 and forms a reference architecture for packet and shared-media oriented communication. Figure 1 compares the OSI-layer model to the IEEE802 reference model [ANS99].

Authentication and authorization are essential parts of every network which sets a high value to security. Especially in the WLAN-domain, these two terms are very important. There are no physical barriers, an unauthorized access to a non-secured Wireless-LAN is easily possible. Everybody who is located in the reception area is theoretically able to communicate with the access point and this circumstance makes data exchange or eavesdropping possible.

2   Extensible Authentication Protocol

EAP replaced earlier authentication methods (PPP, CHAP, etc.). The abbreviation EAP stands for *Extensible Authentication Protocol* and describes a (flexible) extension of PPP. PPP is defined in RFC 2284 (the work for an adapted version, which will be submitted by the IETF, is in progress). It makes the application of different authentication methods (EAP-TLS or TTLS) possible and it is not necessary to publish the method at the LCP phase. In fact the LCP phase only clears the need for an EAP authentication. If an EAP authentication is necessary, further information will be queried from the NAS. EAP can be seen as an open framework for PPP connections and makes the additional usage of further authentication methods possible.

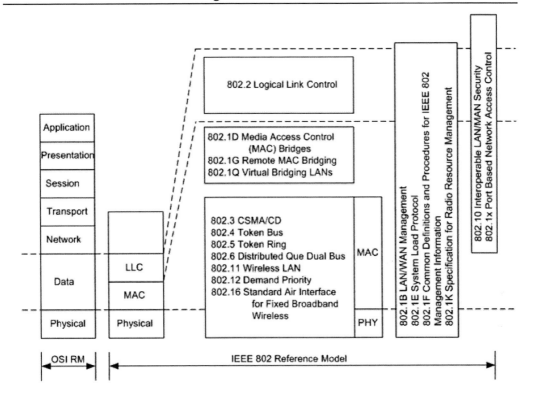

Figure 1: Comparison between OSI reference model and IEEE 802 [SSBM04].

Figure 2 gives an overview of used technologies and methods to apply authentication in the context of EAP. The authentication is based on a three-layer-model, where every participant plays a defined role. The client requests the authentication, the access point acts as the authenticator (controls the access) and the NAS manages user and authentication data. The access point, as authenticator which manages the access to the network, is not in possession of the required information. To decide about a certain network access request, the access point has to forward the request to the NAS. The access point uses the 802.1X protocol and RADIUS respectively to provide each client with a logical port which can be enabled or blocked. IP as transmission protocol is not mandatory. Figure 2 shows, that EAP can also be used with other media than WLAN. Also Ethernet, Token-Ring or the whole 802 family in general are able to use EAP as authentication method. The process of authentication provides the integration of several methods (TLS, CHAP or EAP-TTLS) and shows the flexibility and variety of EAP.

EAP consists of three parts: the messages, the message frames and the extended functionality. As message types 'Request', 'Response', 'Success' and 'Failure' are defined. Figure 3 shows their structure, all other relevant EAP messages and the basic frame format of EAP.

The first byte describes the kind of message, while the identifier takes care of the correct allocation of frames which are transmitted in different sequences. The bytes two and three describe the packet size. The amount of further bytes depends on the used EAP-method and is variable. In the case of a 'Request' or 'Response' frame it is possible that the type

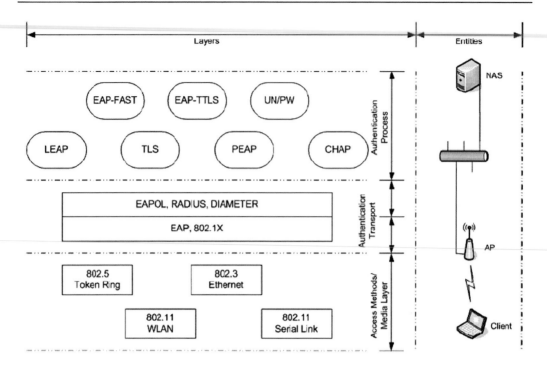

Figure 2: Model of Authentication

identification is bigger than forty values. These values represent the authentication method, which will be used. The number 13, for instance, stands for a EAP-TLS authentication. It has to be considered that the corresponding RFC describes only eight methods; further methods are defined by the IANA (Internet Assigned Numbers Authority).

It is not mandatory that EAP has to be initialized by the access point, also an initialization from the client is possible. EAP is a subset of PPP, which means that an EAP communication is always a point-to-point communication which supports no messages broadcasting. As already mentioned, EAP describes only a framework, where different authentication protocols can be used. Thus RFC 2284 describes only the general process of an EAP communication and leaves specific details open. Specific details are described in other RFCs and so RFC 2284 is only 16 pages long. The listing of all EAP authentication methods is not the scope of this chapter and so Figure 4 demonstrates a concrete use case with EAP-MD5. The functionality of MD5 will be explained in the context of CHAP in [Dav04].

The first message exchange between authenticator and client is an optional process. Although this message type is defined in the EAP RFC, it is noted that the semantic of an identity can not be guaranteed. Thus authentication methods should not trust information exchanged in this phase. The exchange is processed with plain text and so readable for everyone, a clear security vulnerability. Nevertheless EAP-CHAP is using this message sequence to exchange the user name. The user name will subsequent forwarded to the authentication server, which responses a challenge over the authenticator to the client. With the challenge and the password, which was exchanged before, the client computes a hash value and returns it to the server. The server also computes the hash value and com-

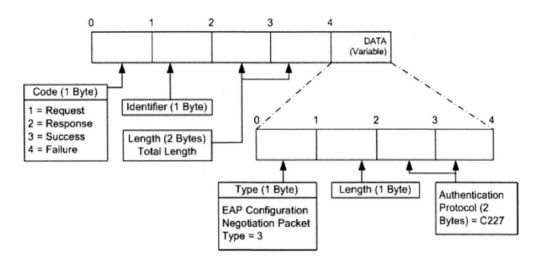

Figure 3: EAP frames

pares both values. If both values are identical, the server sends a success message to the client and the NCP phase can be initialized. In the case of a negative comparison result an initialization of the NCP phase is not possible. After a successful authentication process, further information will be transmitted in plain text. At this point it has to be mentioned that authentication has nothing to do with data encryption. In fact it only has to support the authentication between server and client.

When it is necessary to encrypt data after the authentication phase, further settings have to be considered. In the case of Wireless-LAN structures WEP, WPA or WPA2 with AES would be a possibility. Even if EAP-CHAP is a protocol which is simple to implement, it has, especially in the WLAN-domain, serious security vulnerabilities. EAP-CHAP is susceptible to dictionary- and brute-force-attacks and so Microsoft does not support this authentication protocol in the WLAN-interface of WindowsXP [SSBM04]. Another method is the similar authentication protocol EAP-OTP (One Time Pad7), which is defined in the RFC 2289. It is mainly used in VPN scenarios. EAP-GTC (Generic Token Card) works like EAP-OTP, but with the addition that hardware tokens are used.

2.1 EAP-TLS

The authentication methods presented before, have a clear disadvantage; only the client and not the server will be authenticated. This circumstance makes the installation of rogue APs possible. An attacker who installs such an access point is able to listen on data which is sent to the AP at a client connection attempt. To solve this problem EAP-TLS (Transport Socket Layer), which allows mutual authentication, can be used. With this protocol it is possible that both partners, the client and the server, can be sure that they are communicating with the desired partner or network. Due to the needed PKI (Public Key Infrastructure) the deployment of EAP-TLS is much more work than the deployment of an EAP-CHAP infrastructure. Both sides need digital certificates, which are used to determine/validate the

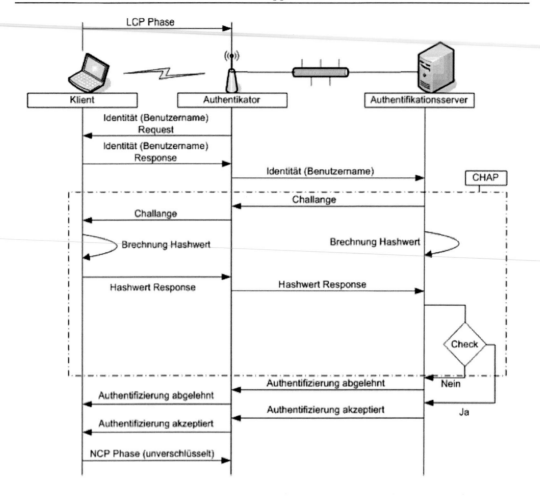

Figure 4: 802.1x

identity of the current communication partner. The deployment of a PKI requires a proper
system maintenance, which is necessary at EAP-TLS. Typical tasks are the certificate administration; the creation, blocking or renewal of expired certificates.

Before EAP-TLS will be discussed in more detail, a general introduction of TLS will
be given. As already mentioned the primary aim of TLS is to ensure data integrity and
confidentiality between the communication partners. Additional TLS is able to provide
mutual authentication. The TLS 'Record Protocol' and the TLS 'Handshake Protocol' are
the main layers of TLS [YM03, 22499]. The Record Protocol is located on the lowest layer,
which build up on a communication protocol like TCP, and provides two features:

- The connection is confidential. This is accomplished by using symmetric encryption
  procedures, like DES or RC4, and corresponding key exchange protocols. Optional
  the record protocol can be used unencrypted.

- The connection is shielded from manipulations, by using integrity based mechanism.
  One way functions, like SHA or MD5 are used to ensure the needed integrity. Although these mechanism are not obligatory for the record protocol to work, their

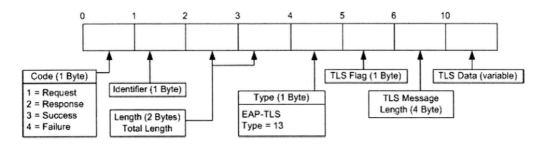

Figure 5: EAP-TLS frame

usage is common practice. The TLS record protocol is used to encapsulate data from network layers above, like the TLS handshake, which provides the following functionalities:

- A both-way authentication of the client and the server.
- The negotiation about which cryptographic algorithm will be used.
- The key exchange.

These processes are all performed before any network layer above sends or receives data.
Following this, TLS provides the following basic functionalities:

- The authentication of both communication partners is achieved by the usage of symmetric or asymmetric cryptography. Although both-way authentication is held optional, a one-way authentication is mandatory for the TLS protocol.
  The negotiation about a shared secret is secure, even if a third unauthorized person is in between the communication partners.
- The negotiation about keys and the used algorithm is secure. No attacker is able to change data without his actions being discovered.
- Another advantage of TLS is the fact that the protocol is transparent, and therefore independent of applications above.

EAP-TLS uses procedures of the TLS protocol, like the TLS handshake with its authentication and negotiation mechanism, to achieve its functionality. Some TLS record protocol methods are not used by EAP-TLS. Figure 4 shows the architecture of an EAP-TLS frame (for this purpose also see figure 3). Figure 6 outlines the EAP-TLS procedure and shows the combination of EAP and TLS .
Consider that the scenario, outlined in figure 5, assumes existing and valid certificates for every communication partner. For convenience, figure 5 does not include the AP, because of its simple task of forwarding data to the server.
First the identity exchange is handled in the initialization phase, which is similar to the EAP-CHAP (see figure 4). Afterwards EAP-TLS is initialized in one frame, which enables the client to trigger the process.
By sending "Client Hello" the client announces himself at the server, which responds by

sending his parameter, respectively certificates. The client himself checks the server's certificate and sends his parameters, respectively certificates, in return.

One important fact is that in WLAN infrastructures EAP-TLS is only used as an authentication mechanism, but not as an encryption mechanism for data. Therefore, depending on the specific implementation of the infrastructure, WEP or WAP is used. Optional WEP or WAP keys can be derived from the EAP-TLS phase.

Following this, the usage of EAP-TLS ensures a valid authentication of both communication partners, but not a valid or secure data transport.

2.2   EAP-TTLS

The IETF draft EAP-TTLS (Tunneled Transport Layer Security) was developed by Funk Software and Certicom, and is currently a working document of the PPP Extension Group. EAP-TTLS is very similar to the EAP-TLS concept, but concentrates only on a server-sided authentication. The client is able to authenticate himself by using other mechanism, like CHAP or Username/Password. TLS is only used as a secure transport tunnel to send data, optional via EAP, to the server.

The encrypted tunnel enables the procedure to resist man-in-the-middle or dictionary attacks. Due to the fact that this method only needs server-sided certificates, the effort spent on the certificate management is reduced. Besides the simplicity of EAP-TTLS another advantage is the compatibility with legacy systems, which are not able to process the EAP protocol. This is achieved by the possible usage of authentication mechanism, which are not based on EAP. A non-technical disadvantage of EAP-TTLS is the poor distribution of WLAN devices using this protocol [Dav04, Sof05].

2.3   PEAP

PEAP (Protected Extensible Authentication Protocol) was developed by Microsoft, Cisco and RSA Security [BSD04] an is at the same time a very secure and a simple solution for WLAN authentication [SSBM04]. PEAP uses, similar to EAP-TTLS, an encrypted TLS tunnel to exchange authentication data between server and client. This compensates weak points of EAP, where some data (e.g. the identity of the communication partners) is sent as plaintext.

Once again the server has to own a valid certificate, in contrast to the client, to enable a successful authentication. Similar to EAP-TTLS the client can use different authentication mechanism.

The PEAP protocol consists of two phases: the preparation of a secure tunnel, via EAP-TLS protocol, and the implementation of the client authentication, based on EAP mechanism. In comparison to EAP-TTLS, PEAP offers the advantage of exchanging selectable and PEAP specific information via the TLS tunnel. The architecture of the PEAP frame is very similar to the EAP-TLS frame, differing only by the type field (PEAP uses 25, EAP uses 13) and by the version-bits in the flag field.

Another mentioned difference to EAP-TTLS is the Type Length Value (TLV), which is used for the transmission of freely selectable data. Due to the fact that this transmission

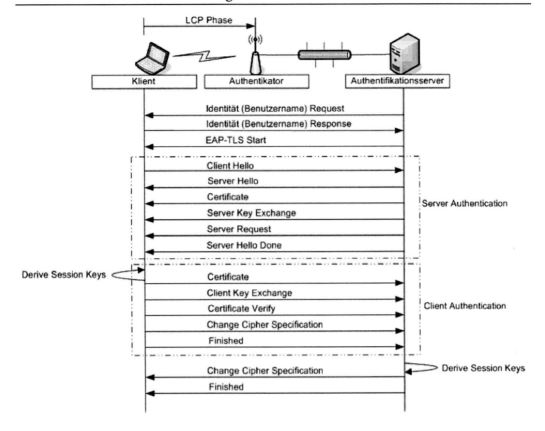

Figure 6: EAP-TLS choreography

takes place in the second phase of the PEAP protocol, the frame format conforms to the type 33 (EAP-TLV). PEAP TLV frames are very flexible in their structure, but depended to specific fabricators.

Due to the circumstance that the RFC itself defines eight different TLV types, PEAP implementations have to understand and be able to process all TLV types [SSBM04]. The choreography of a PEAP communication resembles in most parts a EAP-TLS communication. Similar to EAP-TLS a EAP typical exchange of identity is used in the PEAP protocol, which task it is to identify the server and the client but not to contain confidential authentication data. Those are transmitted afterwards via the secure tunnel.

While EAP-TLS uses a both-way authentication between server and client in the TLS phase, the PEAP protocol's client authentication is executed after the TLS phase [SSBM04, ANS99]. In comparison with EAP-TTLS, the PEAP protocol has the disadvantage that the EAP protocol has to be used whilst the client authentication. A non-technical advantage of PEAP is the wide distribution of WLAN devices using this protocol.

2.4   Comparison of EAP Mechanism

Table 3 recapitulates the three most widely spread EAP mechanism. EAP-TLS and EAP-TTLS provide a bigger application field in non Microsoft operation systems (see table 2). When using EAP-TLS the fact that if the ˝private˝ key is compromised, all client certifi-

	EAP-TLS	EAP-TTLS	PEAP
Status	RFC 2716	Internet draft	Internet draft
SOFTWARE			
Client Implementations	Cisco, Funk, Meetinghouse, MS, Open1x	Funk, Meetinghouse	Microsoft
Supported client plattforms	Linux, Mac OS X, Windows 95/98/ME/NT/XP	Linux, Mac OS X, Windows 95/98/ME/NT/XP	Windows XP
Authentication server implementations by	Cisco, Funk, HP, FreeRADIUS, Meetinghouse, MS	Funk, Meetingshouse	Cisco
Authentication method	Client certificates	Any	Any EAP method
PROTOCOL OPERATIONS			
Basic protocol structure	TLS Session and validate certificates on both sides	(1)TLS Session (2)Exchange of attribute-value pairs	(1)TLS Session (2)EAP exchange over TLS tunnel
Fast session reconnect	No	Yes	Yes
WEP Integration	Server can supply WEP key with external protocol (e.g. RADIUS)		
PKI AND CERTIFICATE PROCESSING			
Server Certificate	Required	Required	Required
Client Certificate	Required	Optional	Optional
Cert Verification	Through certificate chain or OCSP TLS extension (current internet draft		
Effect of private key compromise	Re-issue all server certificates	Re-Issue certificates for servers	
CLIENT AND USER AUTHENTICATION			
Authentication direction	Mutual: digital certs both ways	Mutual: cert for server and tunnelled method for client	Mutual: cert for server and protected EAP for client
Protection of user identity exchange	No	Yes (TLS)	Yes (TLS)

Figure 7: Overview of EAP mechanisms

cates have to be reseted, has to be taken into consideration. This can lead to an enormous administrative effort. Following this, a highly secure PKI has to be provided and the Root CSS has to be offline [Ent05].

3   RADIUS

The RADIUS (Remote Address Dial-In User Service) protocol was developed as an authentication and charging protocol by Livingsten Enterprises. 1996 the RADIUS protocol was given to IETF, and is since then a standardized protocol (RFC 2058,RADIUS Specification and RFC 2059,RADIUS Accounting).

The RADIUS client/server protocol uses UDP, which is a connectionless network protocol and in contrast to TCP does not focus on reliability. The disadvantages of UDP, concerning reliability, are compensated by the RADIUS protocol by using specific intern mechansim.

3.1 UDP versus TCP

The following itemization is provides RADIUS characteristics, which explain why the reliable TCP protocol was not used in the RADIUS protocol:

1. If the request to a primary Authentication server fails, a secondary server must be queried.
   To meet this requirement, a copy of the request must be kept above the transport layer to allow for alternate transmission. This means that retransmission timers are still required.

2. The timing requirements of this particular protocol are significantly different than TCP provides.
   At one extreme, RADIUS does not require a "responsive" detection of lost data. The user is willing to wait several seconds for the authentication to complete. The generally aggressive TCP retransmission (based on average round trip time) is not required, nor is the acknowledgment overhead of TCP.
   At the other extreme, the user is not willing to wait several minutes for authentication. Therefore the reliable delivery of TCP data two minutes later is not useful. The faster use of an alternate server allows the user to gain access before giving up.

3. The stateless nature of this protocol simplifies the use of UDP.
   Clients and servers come and go. Systems are rebooted, or are power cycled independently. Generally this does not cause a problem and with creative timeouts and detection of lost TCP connections, code can be written to handle anomalous events. UDP however completely eliminates any of this special handling. Each client and server can open their UDP transport just once and leave it open through all types of failure events on the network.

Die Abbildung 9 ist im Original falsch referenziert und beschriftet.

3.2 Function Sequence of RADIUS

Figure 8 shows schematically the sequence of actions of a RADIUS session with the participation of an AP as RADIUS client [SSBM04, WCC04]. Prior to this choreography a shared secret must be exchanged between the RADIUS client and server, which is used for data encryption between these two RADIUS participants. In the first step the WLAN client initiates an authorization request to the AP. Thereupon the AP asks the WLAN client for a set of credentials, which can be a combination of username/password or in a packaged format like PPP and EAP respectively. The RADIUS client forwards the credentials and the user-requested port ID to the RADIUS server for authorization. If a password is used, it is encrypted with the MD5 algorithm.

   If the RADIUS server is unreachable, depending on the configuration, the AP can forward the request to another server or retry it again after a specified period of time. When the RADIUS server received the request, it verifies the data against its database. In case the request is not encrypted with the shared secret, it should be discarded. If the validation is successful, the server sends a success message to the AP, which by now is allowed to

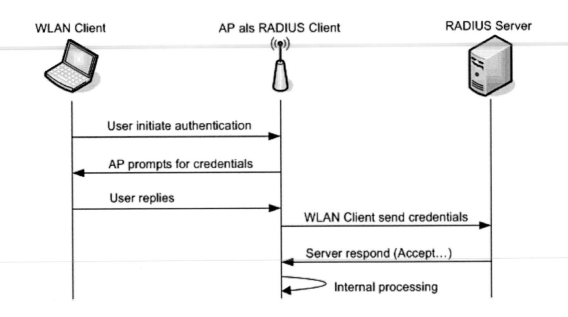

Figure 8: RADIUS choreography

open the requested port. Usually a combination of RADIUS and 802.1X is used in practice. During EAP-based credential exchange, EAP-frames are sent encrypted and as a whole to the server.

3.3   Summary of the RADIUS Functions

The following key functions characterize RADIUS according to RFC2058: Client/server model: a NAS (Network Access Server) operates as client of the RADIUS protocol and is responsible for forwarding the data to and from the RADIUS server. RADIUS servers have the task to process connection requests from clients appropriately, authenticate the user and to provide all necessary connection information (e.g. keys). Thereby the RADIUS server can also act as a proxy client for other RADIUS services.

3.4   IEEE 802.1X

IEEE 802 LANs are often implemented in environments (e.g. company networks or campus networks), which make it possible for unauthorized devices to establish a physical connection to the network - without permission or via network plugs open to the public. In such an environment it is desirable to regularize the access to services, provided by the network, in such a way that it is impossible for an attacker to gain access without authentication and authorization. For this purpose the IEEE 802.1X standard defines a mechanism for port based network access control for miscellaneous 802 LANs [CJL05]. 802.1X provides an authentication framework where the user is authenticated by a central authority (RADIUS). Authentication is done using EAP types.

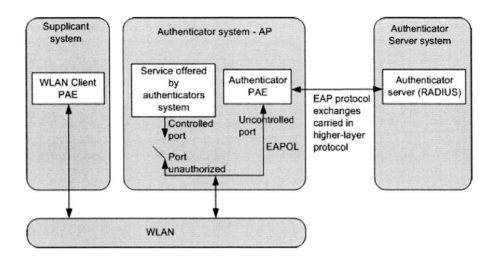

Figure 9: IEEE 802.1X framework for a WLAN environment

Albeit RADIUS is only within 802.1X optimal, in practice an 802.1X authenticator is usually a RADIUS client as well. Especially in connection with WLAN, RADIUS and 802.1X are tight coupled. 802.1X does not depend on a server in the background to carry out its functionality; it can be implemented autonomous into network devices (AP, Bridges, etc.). In 802.1X the port is the logical connection between the authenticator (in the case of WLAN the AP), the requesting WLAN client and possibly an authentication server in the background, which mostly operates as RADIUS server. The following figure 9 shows the IEEE 802.1X framework for a WLAN environment schematically. That follows a discussion on the involved components. As can be seen in figure 9, the supplicant system and the authenticator system have a PAE (Port Access Entity), which is responsible for processing the algorithms and protocols of the respective authentication mechanism. The authenticator PAE manages the state of the controlled and uncontrolled port which can be, dependent on the authentication result, closed or activated (in the figure the controlled port is closed, which is illustrated by the broken link). As long as the WLAN client is not authenticated it can only access and respectively use the uncontrolled port. Thereby it is guaranteed that an attacker cannot access services offered by the authenticator system without being authenticated by the authentication server. Only after successful authentication the authenticator PAE opens the port (visually spoken, closing the link to the controlled port in figure 9). Now the WLAN client is able to access the desired services.

It is important to understand that these ports are logical ports and do not exist physically. 802.1X applies mainly EAP (IETF RFC 2284) which allows for offering various authentication mechanisms, including MD5 (IETF RFC 1321), TLS, PEAP and lots of others. Additionally 802.1X defines EAPOL (EAP over LAN). As mentioned previously, in the WLAN field 802.1X and RADIUS mostly coexist to take advantage of both mechanisms and consequently to establish a high level of security. For this purpose the authentication system (the AP) embeds the received EAP messages from the requesting WLAN client into RADIUS packages and transmits them to the authentication server which is implemented as RADIUS server (see RFC 2869, RADIUS Extensions). Thus the 802.1X protocol offers effectively managed access control, independent of chosen authentication and encryption methods.

Furthermore it would be possible to raise the access protection by additionally implementing a MAC based control system. The drawback of this decision is a significant increase of the administrative expenses. Optionally, if session keys are used, the 802.1X protocol can send them in the accept message back to the requesting client. 802.1X only offers access control but no authentication mechanisms, hence to use it an appropriate authentication mechanism, such as EAP, has to be chosen.

4   RADIUS in a Windows Environment

4.1   IAS - Internet Authentication Server

In a WLAN infrastructure the IAS (Internet Authentication Server) acts as a RADIUS server and processes the authentication requests from clients which were forwarded by the AP. The IAS offers a centralized contact point for authentication and also for accounting information. In principle it does not matter if the authentication is carried out over WLAN, VPN, Dial-Up or between routers because IAS supports a multitude of methods (including PAP, CHAP, EAP, etc.). In order to offer the RADIUS functionality, the IAS server implements the IETF standards RFC 2865 and 2866, whereby it is possible to run the ISA server as RADIUS proxy. In case Active Directory is installed, the IAS server will use it as authentication information database. Depending on the Windows Server 2003 edition, the ISA server supports up to 250 RADIUS clients. A RADIUS Windows environment is shown in figure 10.

4.2   Network Security

The communication between the client and the RADIUS server is authenticated by a shared secret, which ideally is never sent over the network. Rather the shared secret should be set manually on every client and server. In addition every user-password is sent encrypted to avoid password theft through wiretapping.

4.3   Flexible Authentication Mechanisms

A RADIUS server supports a multitude of methods to authenticate a user (e.g. PPP, PAP, UNIX Login or CHAP).

4.4   Extensible Protocol

The RADIUS protocol allows flexible extensions by implementing variable three-tuple-attributes. In the future new mechanisms can be added without an implementation change.

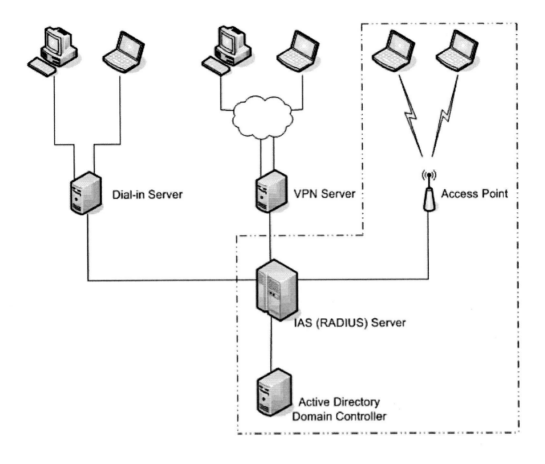

Figure 10: RADIUS in a Windows environment

5   Conlusion

In this chapter we explained how RADIUS can be used to secure a wireless LAN. The authentication of a user is performed by an authentication server and only after having been correctly authenticated is the user permitted full access to the local area network. In contrast to WEP authentication, RADIUS authenticates users and not only machines; moreover the security-relevant operations are mainly handled by a separate server and not in the access point itself.

References

[22499]   RFC 2246. The tls protocol version 1.0, 01 1999.

[ANS99]   ANSI/IEEE. Standard 802.11 (r2003), 1999.

[BSD04]   Bhagyavati, Wayne C. Summers, and Anthony DeJoie. Wireless security tech-
           niques: An overview. *ACM Press*, pages 82–87, 10 2004.

[CJL05]   Jyh-Cheng Chen, Ming-Chia Jiang, and Yi-Wen Liu. Wireless lan security and ieee 802.11i. *IEEE Wireless Communications*, pages 27–36, 02 2005.

[Dav04]   Joseph Davies. *Deploying Secure 802.11 Wireless Networks with Microsoft Windows*. Microsoft Press, 2004.

[Ent05]   Symbol Enterprise. The use of digital certificates for authentication to a wireless lan. Technical white paper, Symbol Enterprise, 02 2005.

[Sof05]   Funk Software. www.funk.com. WEB, 10 2005.

[SSBM04]   Krishna Sankar, Sri Sundaralingam, ANdrew Balinks, and Darrin Miller. *Cisco Wireless LAN Security, Expert Guidance for securing your 802.11 Networks*. Cisco Press, 2004.

[WCC04]   R.C. Wang, R.Y. Chen, and Han-Chieh Chao. Aaa architecture for mobile ipv6 based on wlan. *International Journal of Network Management*, pages 305–313, 07 2004.

[YM03]   Xiuying Cao Yue Ma. How to use eap-tls authnetication in pwlan enviroment. *IEEE International Conference Neural Networks and Signal Procesing*, 12 2003.

In: Mobile Multimedia: Communication Engineering …   ISBN: 1-60021-207-7
Editors: I.K. Ibrahim and D. Taniar pp. 117-135   © 2006 Nova Science Publishers, Inc.

Chapter 7

REAL-TIME VIDEO TRANSPORT
IN HETEROGENEOUS IP NETWORKS

Panagiotis Papadimitriou*, Vassilis Tsaoussidis**

Electrical & Computer Engineering Department, Demokritos University,
Xanthi, 67100, Greece

Abstract

We assess the performance of real-time video delivery within the context of transport protocol support and efficiency. We mainly focus on real-time QoS management in heterogeneous wired/wireless networks. We investigate real-time video requirements versus the QoS provided by the underlying network. Furthermore, we analyze some representative congestion control mechanisms, which are suitable for preserving the fundamental QoS guarantees for time-sensitive traffic. Based on simulation results, we quantify video delivery reaching several conclusions on TCP/UDP efficiency in such environments. Finally, we identify the mechanisms that manage to alleviate most of the impairments caused by the network heterogeneity.

1 Introduction

Multimedia flows carrying audio and video information constitute a considerable portion of Internet traffic. Time-sensitive applications yield satisfactory performance only under certain *Quality of Service* (*QoS*) provisions, which may vary depending on the application task and the type of media involved. Streaming video, in particular, is comparatively intolerant to delay and variations of throughput and delay. Furthermore, video quality is commonly impaired by reliability factors, such as packet drops due to congestion or link errors.

Such stringent requirements necessitate explicit management techniques in order to preserve the fundamental QoS guarantees for video traffic. In this context, *Internet Engineering Task Force* (*IETF*) attempted to facilitate true end-to-end QoS on IP networks by

---

* E-mail address: ppapadim@ee.duth.gr
** E-mail address: vtsaousi@ee.duth.gr

defining *Integrated* (*IntServ*) and *Differentiated Services* (*DiffServ*) models [16, 3]. IntServ follows the signaled-QoS model, where the end-hosts signal their QoS need to the network, while DiffServ works on the provisioned-QoS model, where network elements are set up to service multiple classes of traffic with varying QoS requirements. However, both models are associated with high implementation costs and limited applicability; hence, they have not yet received wide appeal from the majority of users. In reality, most end-users still rely on the best-effort services of the Internet which strives to meet the high demands of the emerging multimedia applications. Since the Internet provides services for various types of applications, flow contention inevitably appears in varying traffic, spatial and temporal patterns. In the event of limited bandwidth availability, coexisting traffic (e.g. web, FTP) commonly impacts the timely delivery of video-packets degrading application performance. Inversely, video flows may also hurt congestion-sensitive traffic when they compete for scarce bandwidth. Although the mixture of video with data traffic may call for complex admission policies, in this study we assume that real-time video does not get any prioritization.

Transmission Control Protocol (*TCP*) is basically designed to provide a reliable service for wired Internet. The *Additive Increase Multiplicative Decrease* (*AIMD*) algorithm [5], incorporated in standard TCP versions, achieves stability and converges to fairness when the demand of competing flows exceeds the channel bandwidth. TCP is further enhanced with a series of mechanisms for congestion management including *Congestion Avoidance* [10], *Slow Start*, *Fast Retransmit* and *Fast Recovery* [17]. Despite these features, TCP demonstrates inadequate performance in heterogeneous wired/wireless environments. Authors in [18] outline three major shortfalls of TCP: (i) ineffective bandwidth utilization, (ii) unnecessary congestion-oriented responses to wireless link errors (e.g. fading channels) and operations (e.g. handoffs), and (iii) wasteful window adjustments over asymmetric, low-bandwidth reverse paths. More precisely, a suitable TCP for wired/wireless networks should be able to detect the nature of the errors that result in packet losses in order to determine the appropriate error-recovery strategy. Based on such an approach, the sender would not be obliged to reduce its transmission rate in the event of a wireless error or handoff. A next level of enhancement for TCP would enable a more sophisticated error-recovery strategy adjusted to the error characteristics of the underlying network, device constraints and performance tradeoffs.

The difficulty of the task that TCP has to perform is further enhanced, when the protocol provides services for time-sensitive applications. Standard TCP usually induces oscillations in the achievable transmission rate and occasionally introduces arbitrary delays, since it enforces reliability and in-order delivery. Evaluating the performance of real-time traffic based on traditional performance metrics (e.g. throughput) may produce misleading results, since such metrics do not account for variable delays that usually degrade real-time applications. Consequently, the inefficiency of TCP in conjunction with the lack of accurate performance metrics render the applicability of TCP for real-time applications limited.

In this context, several TCP protocol extensions [7, 2] have emerged to overcome the standard TCP limitations providing more efficient bandwidth utilization and sophisticated mechanisms for congestion control, which preserve the fundamental QoS guarantees for time-sensitive traffic. *TCP-friendly* protocols, proposed in [7, 21, 22] achieve smooth window adjustments, while they manage to compete fairly with TCP flows. In order to achieve smoothness, they use gentle backward adjustments upon congestion. However, this modification has a negative impact on protocol responsiveness [20]. Binomial schemes [2], such as *IIAD* or *SQRT* [6], are quite attractive to multimedia applications for their smooth rate variations. However, they are not able to achieve TCP-friendliness independent of link capacity [4].

User Datagram Protocol (*UDP*) has been widely used instead of TCP in real-time applications. UDP lacks all basic mechanisms for error recovery and flow/congestion control. Thus, it allows for transmission attempts at application rate. That said, UDP can not guarantee reliability, and certainly is not able to deal with network delays either. In [13] we have shown that UDP may perform worse than TCP in several occasions.

Our objective is to explore the performance of the most prominent end-to-end solutions in heterogeneous environments from the perspective of real-time video delivery. End-to-end schemes are promising, since significant performance gains can be achieved without any extensive support from intermediate nodes in the network. More precisely, we evaluate UDP versus TCP, investigating whether UDP exhibits superior performance, as implied. Beyond UDP and TCP supportive role, we focus on selected transport layer mechanisms that provide a viable alternative for unicast real-time flows (instead of allowing a free-transmitting policy). "Blind" congestion control schemes that rely on specific events triggered by violated thresholds (e.g. AIMD) are generally discouraged for time-sensitive traffic [13]. Furthermore, they exhibit an inherent weakness in heterogeneous wired/wireless networks, since they sharply reduce the sending rate in response to each loss event. Therefore, we study TCP mechanisms that dynamically exploit bandwidth availability relying on precise measurements of the prevailing network conditions. In this context, equation-based congestion control, as well as a measurement-based approach are under investigation. Through extensive simulations, we address the challenges and implications induced by network heterogeneity, such as various types of competing traffic, multiple bottlenecks, diverse *Round Trip Times* (*RTTs*) and transient errors, and further study the associated impact on perceived video quality.

We organize the rest of the paper as follows. In the following section, we describe the particularity of real-time video traffic in terms of characteristics and requirements. In Section 3, we analyze selected end-to-end congestion control mechanisms, which are highly advisable for time-sensitive traffic. Section 4 includes our evaluation methodology, followed by Section 5, where we discuss the results of our experiments. Finally, in the last section we highlight our conclusions.

2   Real-Time Video QoS Parameters

The task of specifying the effects of network QoS parameters on video quality is challenging. Transmission rate fluctuations, increased delays, jitter and packet loss commonly deteriorate the perceived quality or fidelity of the received video content. In the sequel, we discuss the effects of end-to-end QoS parameters on real-time video delivery. Although we refer to these parameters individually, we note that they do not affect quality in an independent manner; they rather act in combination or cumulatively, and ultimately only this joint effect is detected by the end-user. However, studying the effects of network parameters in an isolated fashion is a more tractable approach.

2.1 End-to-end Delays

End-to-end delay consists of the delay incurred by the video signal from the instant it is produced by the video source, until its playback at the receiver. Initially, the signal is encoded, followed by the packetization phase, incurring an encoding (D_{enc}) and packetization (D_{pack}) delay, respectively. Video packets are then transmitted on the network. Network delay is expressed by the summation of propagation (P_h), transmission (T_h), and the variable queuing and processing delays (Q_h) for each hop h in the path from the source to the destination. If we include a playback delay (D_{play}) and we ignore the processing delays at both sender and receiver, the end-to-end delay D for a packet is expressed, as:

$$D = D_{enc} + D_{pack} + \sum_{h \in Path}(T_h + Q_h + P_h) + D_{play} \qquad (1)$$

We observe than for a certain video codec and network connection the only random component in equation (1) is queuing delay. Therefore, end-to-end latency is mostly affected by queuing delays.

The typical delay guidelines for real-time video are depicted in Table 1. End-to-end delays exceeding 300 ms affect the timely delivery of video data and have a negative impact on perceived quality.

Table 1. Typical delay guidelines for real-time video

Delay	Effect in perceived quality
Less than 150 ms	Delay is not noticeable
150 - 250 ms	Acceptable quality with slight visual impairments
Over 250 - 300 ms	Video quality is degraded

2.2 Delay Variation

Delay variation is usually caused by the variable queuing and processing delays on routers during periods of increased traffic and occasionally by routing changes. Delay variation is responsible for the phenomenon called network jitter, which has unpleasant effects in real-time communication, as packets often reach the receiver later than required. Practically, delayed packets are either discarded and considered lost, or at the worst they obstruct the proper reconstruction of oncoming packets. Table 2 depicts the perceptual effect of jitter on real-time video. Generally, time-sensitive traffic does not afford delay variations above 75 ms. Excessive jitter usually generates jerky video playback, which is intolerable by most users.

Table 2. Typical jitter guidelines for real-time video

Delay Variation	Effect in perceived quality
Less than 40 ms	Jitter is not noticeable
40 - 75 ms	Acceptable quality with minor impairments
Over 75 ms	Unacceptable quality, too much jitter

2.3 Packet Loss

Packet loss composes an impairment factor, since it causes a perceptible degradation in video quality. The specific impact is particularly disruptive on compressed video streams, since packet drops induce distortions on the visual quality, which are typically more annoying to the human viewer than most types of impairments (e.g. encoding artifacts). Packet loss is typically the result of excessive congestion in the network. However, in heterogeneous wired/wireless environments, apart from congestion, hand-offs and fading channels may result in packet loss. Standard TCP is unable to successfully detect the nature of the errors in such a network environment. As a result, TCP is not able to determine the appropriate error-recovery strategy with a negative impact on the performance of video delivery. Numerous systems use error correction codes (*ECC*), such as *Forward Error Correction* (*FEC*), in order to ameliorate the impact of packet loss on visual quality. Despite the presence of such mechanisms, the perceived quality is inevitably diminished, when the transport protocol is unable to restrict packet drops. Furthermore, the use of FEC or retransmission introduces additional delays that delay-sensitive flows cannot withstand.

In [14] Paxson investigated the loss rate for a number of Internet paths and found that it ranged from 0.6% to 5.2%. A recent study [1] confirmed Paxson's earlier results, but also showed that the average loss rate for the measurements was 0.42%. However, due to an average burstiness of 72%, packet loss can be occasionally much higher and therefore, efficient transport services are required in order to deliver a regular and smooth video flow.

3 Congestion Control Schemes for Time-Sensitive Traffic

In the sequel, we discuss some representative end-to-end congestion control schemes, which are suitable for preserving the fundamental QoS guarantees for time-sensitive traffic. We focus on transport layer mechanisms, since an overview of Internet's current congestion control paradigm reveals that routers play a relatively passive role: they merely indicate congestion through packet drops or *Explicit Congestion Notification* (*ECN*) [15]. It is the end-systems that perform the crucial role of responding appropriately to these congestion signals. Along these lines, our study does not account on QoS functionality in routers, such as *Random Early Drop* (*RED*) [8], *ECN* or other *Active Queue Management* (*AQM*) mechanisms.

3.1 TCP-Friendly Congestion Control

The differences between standard TCP and TCP-friendly congestion control lie mainly in the specific values of additive increase rate α and multiplicative decrease β; their similarities lie in their AIMD based congestion control (a characteristic that enables us to include them both in the family of TCP (α, β) protocols). Standard TCP is therefore viewed as a specific case of TCP (α, β) with $\alpha = 1$ and $\beta = 0.5$. On the other hand, TCP-friendly protocols are designed to satisfy the requirements of time-sensitive applications. However, they may exhibit further weaknesses, when bandwidth becomes available rapidly [20]. Apparently, the tradeoff between responsiveness and smoothness can be controlled to favor some applications, but it may cause some other damages. The choice of parameters α and β has a direct impact on the responsiveness of the protocols to conditions of increasing contention or bandwidth availability.

Several studies are focused on the quantitative modeling of the TCP long-term throughput, as a function of the measured packet loss rate (p), RTT, and initial transmission timeout value (T_0). According to [22], TCP(α, β) throughput can be modeled as:

$$T_{\alpha,\beta}(p,RTT,T_0,b) = \frac{1}{RTT\sqrt{\frac{2b(1-\beta)}{\alpha(1+\beta)}}\,p + T_0\,\min\left(1,3\sqrt{\frac{(1-\beta^2)b}{2\alpha}}\,p\right)p(1+32p^2)} \tag{2}$$

where b is the number of packets acknowledged and α, β are the congestion control parameters. Observations of the window dynamics and event losses are frequently assumed within a time period of a *congestion epoch*. A congestion epoch is defined in [19] as the time period that reflects the *"uninterrupted growing lifetime of a window"*. More precisely, a congestion epoch begins with βW packets, increased by α packets per *RTT* and reaching a congestion window (*cwnd*) of W packets, when a packet is dropped. The *cwnd* is then decreased to βW. Hence, a congestion epoch involves:

$$n = (\beta / \alpha * W + 1) \text{ RTTs} \tag{3}$$

TCP-friendly (α, β) protocols approximate the throughput of standard TCP (α=1, β=0.5), which means that equation (4), which is derived from equation (2), provides a rough guide to achieve friendliness:

$$T_{\alpha,\beta}(p, RTT, T_0, b) = T_{1,0.5}(p, RTT, T_0, b) \tag{4}$$

However, since the network or application conditions may change rapidly, friendliness might not be attained. More precisely, based on equation (3) we conclude that equation (4) can be achieved later than expected, since multiple drops extend further the time of convergence. Furthermore, according to equation (3), the time period required for equation (4) to hold is in reverse proportion to the number of flows within a fixed bandwidth channel; the smaller the number, the larger the window.

TCP-Friendly Rate Control (TFRC) [7] is a representative TCP-friendly protocol, which adjusts its transmission rate in response to the level of congestion, as estimated based on the calculated loss rate. Multiple packet drops in the same RTT are considered as a single loss event by TFRC and hence, the protocol follows a more gentle congestion control strategy. More precisely, with respect to equation (2), the TFRC sender uses the following TCP response function:

$$T(p, RTT, RTO) = \frac{1}{RTT\sqrt{\frac{2p}{3}} + RTO\left(3\sqrt{\frac{3p}{8}}\right)p(1+32p^2)} \tag{5}$$

where p is the steady-state loss event rate and RTO is the retransmission timeout value. Equation (5) enforces an upper bound on the sending rate T. According to [7], TFRC's increase rate, which is solely determined by the value of α, never exceeds 0.14 packets per RTT (or 0.28 packets per RTT when history discounting has been invoked). In addition, the protocol requires 5 RTTs in order to halve its sending rate. Consequently, the instantaneous throughput of TFRC has a much lower variation over time. TFRC eventually achieves the smoothing of the transmission gaps and therefore, is suitable for applications requiring a smooth sending rate. However, this smoothness has a negative impact, as the protocol becomes less responsive to bandwidth availability [20]. TFRC has another major constraint: it is designed for applications transmitting fixed sized packets, and consequently its congestion control is unsuitable for applications that use packets with variable size. In order to overcome this inconvenience, a TFRC variant, called *TFRC-PacketSize* (*TFRC-PS*), has been alternatively proposed.

TCP-Real [19, 23] is a high-throughput transport protocol that incorporates congestion avoidance mechanism in order to minimize transmission-rate gaps. As a result, the protocol is suited for real-time applications, since it enables better performance and reasonable playback timers. TCP-Real approximates a receiver-oriented approach beyond the balancing trade of the parameters of additive increase and multiplicative decrease. The protocol introduces another parameter, namely γ, which determines the window adjustments during congestion avoidance. More precisely, the receiver measures the data-receiving rate and attaches the result to its acknowledgments (*ACKs*), directing the transmission rate of the sender. When new data is acknowledged and *cwnd* is adjusted, the current data-receiving rate is compared against the previous one. If there is no receiving rate decrease, *cwnd* is increased by 1 *Maximum Segment Size* (*MSS*) every *RTT* ($\alpha = 1$). If the magnitude of the decrease is small, the *cwnd* remains temporarily unaffected; otherwise, the sender reduces the *cwnd* multiplicatively by γ. In [19] a default value of $\gamma = 1/8$ is suggested. However, this parameter can be adaptive to the detected conditions. Generally, TCP-Real can be viewed as a TCP (α, β, γ) protocol where γ captures the protocol's behavior prior to congestion, when congestion boosts up.

3.2   TCP-Westwood Bandwidth Estimation Scheme

TCP Westwood [11] is a TCP-friendly protocol that emerged as a sender-side-only modification of TCP Reno congestion control. TCP Westwood exploits end-to-end bandwidth estimation in order to adjust the values of slow-start threshold (*ssthresh*) and *cwnd* after a congestion episode. The protocol incorporates a recovery mechanism which avoids the blind halving of the sending rate of TCP Reno after packet losses and enables TCP Westwood to achieve a high link-utilization in the presence of wireless errors. The specific mechanism considers the sequence of bandwidth samples *sample_BWE[n]* obtained using the *ACKs* arrivals and evaluates a smoothed value, *BWE[n]*, by low-pass filtering the sequence of samples, as described by the following pseudocode:

Algorithm 1. TCP-Westwood

---

if an ACK is received
    sample_BWE[n] = (acked * pkt_size * 8) / (now - last_ACK_time)
    BWE[n] = (1 - beta) * (sample_BWE[n] + sample_BWE[n - 1]) / 2 + beta * BWE[n - 1]
    end if

---

where *acked* is the number of segments acknowledged by the last *ACK*; *pkt_size* is the segment size in bytes; *now* is the current time; *last_ACK_time* is the time the previous *ACK* was received; *beta* is the pole used for the filtering (a value of 19/21 is suggested). However, in [12] we showed that TCP Westwood tends to overestimate the available bandwidth, due to ACKs clustering. *TCP Westwood+* is a recent extension of TCP Westwood, based on the *Additive Increase/Adaptive Decrease* (*AIAD*) mechanism. Unlike the initial version of Westwood, TCP Westwood+ computes one sample of available bandwidth every RTT using all packets acked in the specific RTT, therefore obtaining more accurate estimates [9].

4 Evaluation Methodology

4.1 Scenarios and Parameters

The evaluation plan was implemented on the *NS-2* network simulator. In order to assess the performance on video delivery, we implemented an experimental MPEG-4 video streaming server. The traffic generated closely matches the statistical characteristics of an original MPEG-4 video trace. We developed three separate *Transform Expand Sample* (*TES*) models for I, P and B frames, respectively. The resulting video stream is generated by interleaving data obtained by the three models.

Simulations were initially conducted on the typical single-bottleneck *dumbbell* topology with a bottleneck capacity of 10 Mbps and a round-trip link delay of 64 ms (Fig. 1). We also enabled simulations on a complex network topology (Fig. 2), which addresses the heterogeneity of the Internet. The specific topology includes multiple bottlenecks, cross traffic and wireless links. The propagation delays of the access links from all the source nodes, as well as the links to the peripheral sink nodes range from 5 ms to 15 ms, while the corresponding bandwidth capacities range from 2 Mbps to 10 Mbps. The router *R1* is the bottleneck for MPEG traffic (flows between main source and sink nodes), while the router *R2* is another bottleneck for competing MPEG and cross traffic (flows between peripheral source and sink nodes). Cross traffic includes various FTP flows over TCP Reno: a common application running over a widely used transport protocol. The number of source and sink nodes is always equal. We also attached an equal number of peripheral source and sink nodes.

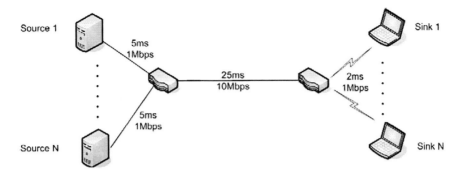

Figure 1. Dumbbell Topology

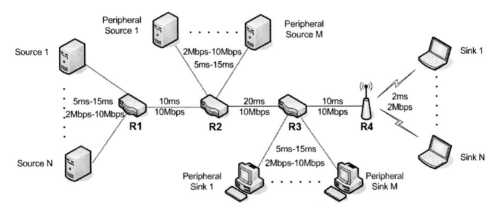

Figure 2. Cross-traffic Topology

NS-2 error models were inserted into the access links to the sink nodes for both topologies. Error models were configured on both (forward and reverse) directions of the link traffic. We used the *Bernoulli* model in order to simulate the link-level errors with configurable packet error rate (*PER*) adjustments. In our experiments, we adjusted PER at 0.01, unless otherwise explicitly stated. In both topologies we used drop-tail routers with buffer size adjusted in accordance with the bandwidth-delay product. Furthermore, we set the packet size to 1000 bytes and the maximum congestion window to 64 KB for all TCP connections. Each simulation lasts for 60 seconds, and diverse randomization seeds were used in order to reduce simulation dynamics. Besides UDP, we focus on the evaluation of the rate-based TFRC and the measurement-based TCP Westwood+ (TCPW+), since both are implied to yield remarkable efficiency over a wide range of network and session dynamics.

4.2   Measuring Performance

We hereby refer to the performance metrics supported by our simulation model. In the cross-traffic network topology, our performance metrics are applied separately to main MPEG and FTP cross traffic. *System goodput* is used to measure the overall system efficiency in bandwidth utilization. System goodput is defined as:

$$Goodput = Original\_Data \, / \, Connection\_time$$

where *Original_Data* is the number of bytes delivered to the high-level protocol at the receiver (i.e. excluding retransmitted packets and overhead) and *Connection_time* is the amount of time required for the data delivery.

Fairness is measured by the *Fairness Index*, derived from the formula given in [5], and defined as:

$$\text{Fairness Index} = \frac{(\sum_{i=1}^{n} \text{Throughput}_i)^2}{n(\sum_{i=1}^{n} \text{Throughput}_i^2)}$$

where *Throughput$_i$* is the throughput of the ith flow and n is the total number of flows.

In order to quantify the performance on video delivery, we monitor packet inter-arrival times and eventually distinguish the packets that can be effectively used by the client application from delayed packets (according to a configurable inter-arrival threshold). The proportion of delayed packets is reflected in *Delayed Packets Rate*. Along these lines, each recipient, receiving packets from the MPEG streaming application, calculates the number of *delayed packets* based on the following algorithm:

Algorithm 2. Delayed Packets

\# For each packet received with sequence number i, determine
\# whether it is delayed

if threshold > 0 **then**
    set packetTime[i] = currentTime
    increase packetsReceived
    if i > 0 **and** packetTime[i] - PacketTime[i - 1] > threshold **then**
        increase delayedPackets
    end if
end if

Several notations used in the pseudocode algorithms are as follows:

1.   *threshold*   : packet inter-arrival time threshold
2.   *delayedPackets*   : number of packets with inter-arrival times exceeding the threshold
3.   *packetTime*   : packet arrival time

In accordance with video streaming requirements, we adjusted the inter-arrival threshold at 75 ms. Since real-time traffic is sensitive to packet losses, we additionally define *Packet Loss Rate*, as the ratio of the number of lost packets over the number of packets sent by the application.

5   Results and Discussion

In the sequel, we demonstrate and comment on the most prominent results from the experiments we performed based on three distinct scenarios. The basic parameters of each simulation scenario are as described in the previous section.

5.1   Performance on Video Delivery

Initially, we conducted a series of experiments on the dumbbell topology in order to evaluate the video performance delivered by the selected protocols. We simulated a wide range of MPEG flows (1-50) adjusting the contention accordingly. The specific scenario features TCP/UDP connections exhibiting approximately the same RTTs. Along these lines, we measured goodput rate, and we additionally selected statistics from delayed packets, since they compose an influencing factor for perceived video quality. The associated results are depicted in Figs. 3-4. Furthermore, we present traces of the queue-length of the bottleneck router (Fig. 5) in the situation of 40 MPEG flows for each protocol separately.

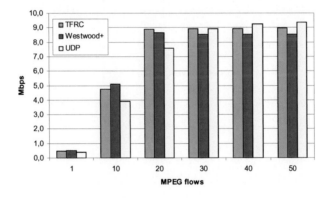

Figure 3. System Goodput

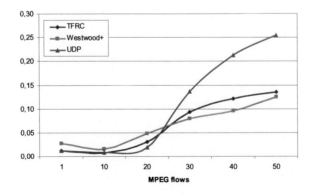

Figure 4. Delayed Packets Rate

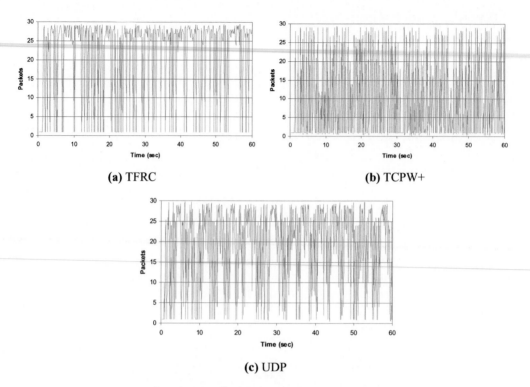

Figure 5. Bottleneck Queue Length (40 flows)

Table 3. Average Bottleneck Queue Length (40 flows)

Protocol	Average Queue Length
TFRC	21.4
TCP Westwood+	10.1
UDP	18.8

Since UDP does not incorporate any close-looped control mechanism, the sender keeps transmitting at application rate. As a result, in the situation of increased link multiplexing (30-50 flows), the protocol exhibits high bandwidth utilization outperforming the rest of the protocols (Fig. 3). Despite the gains in goodput, the unresponsive UDP does not yield satisfactory performance on video delivery. Fig. 4 illustrates that a significant proportion of MPEG packets reaching the recipient exceed the delay requirements of real-time video. The significant delays in video-data delivery are primarily due to the increased queuing delays. Based on equation (1), we showed that queuing delay is a critical factor for the end-to-end latency of real-time video. Fig. 5c, as well as Table 3 demonstrate that UDP results in rapidly growing queues, and eventually in bottleneck buffer overflows. Since the protocol never reduces its transmission rate, the buffer remains overflowed for a considerable amount of time and a significant proportion of the packets that are not dropped, reach the recipient later than required. Therefore, UDP results in long and variable delays, which usually cause transmission gaps and generally degrade the video quality with frustrating consequences to the end-user.

On the other hand, the TCP protocols demonstrate a different behavior. More precisely, with the appropriate selection of increase/decrease parameters both protocols manage to effectively utilize the available bandwidth in the situation of low contention (flows 1-20). In the specific case, Fig. 3 depicts that TFRC, as well as TCPW+ outperform the unresponsive UDP. Furthermore, we note that both protocols achieve high link utilization, as contention increases, despite the unforeseen congestion incidents. Unlike UDP, TFRC and TCPW+ are able to change the sending rate adaptively, although in a different fashion. The incorporated mechanisms for smooth rate control enforce sufficient real-time performance, regardless of link multiplexing (Fig. 4). The proportion of delayed packets is acceptable at most cases and hence, the video stream can be effectively reconstructed at the receiver. Such a performance does not necessitate the use of deep playback buffers in order to ameliorate the effect of jitter.

Comparing the performance on video delivery between the TCP variants, we point out a slight deficiency of TFRC, which is depicted in Fig. 4. TFRC is designed to respond to a loss event (which may include several packet drops) instead of a packet loss. However, in environments with random transient errors, TFRC occasionally fails to obtain accurate estimates of the loss event rate, invoking an inappropriate equation-based recovery that impacts its performance. Furthermore, TFRC's gentle backward adjustments in response to congestion may favor smooth video delivery in normal conditions, but occasionally fail to relinquish a considerable amount of the allocated network resources during congestion periods. As a result, queues in the bottleneck buffer are rapidly built up (Fig. 5a). A comparative view at Figs. 5(a, b) reveals a tendency of TFRC to buffer overflows. Inversely, TCPW+ measuring the rate of ACKs stream performs sharper backward adjustments, enforcing a more persistent buffer draining phase (Fig. 5b). Besides the reported results, the average queue length measurements, as depicted in Table 3, justify our outcome.

5.2   Performance with Heterogeneous Networks

Departing from a comparative overview of protocol performance in terms of video delivery, we study the specific impact over highly heterogeneous environments featuring cross traffic, multiple bottlenecks, diverse RTTs and wireless links. Along these lines, we carried out a series of similar experiments on the cross-traffic topology, which inherently addresses the afore-mentioned features. We simulated a wide range of MPEG flows (1-50) competing with 10 FTP connections running over TCP Reno. The target sending rate for each MPEG flow is adjusted at 640 Kbps in order to enforce strong contention with interfering FTP traffic. The associated results appear in Figs. 6-8. In addition, we present traces of the queue-length of the router R3 (Fig. 9) in the presence of 20 MPEG flows (plus 10 FTP connections).

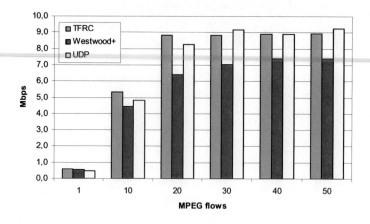

Figure 6. Goodput of MPEG flows

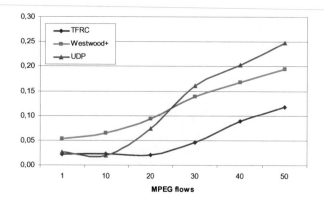

Figure 7. Delayed Packets Rate

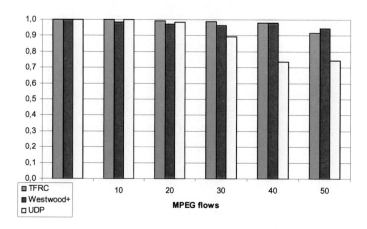

Figure 8. Fairness Index

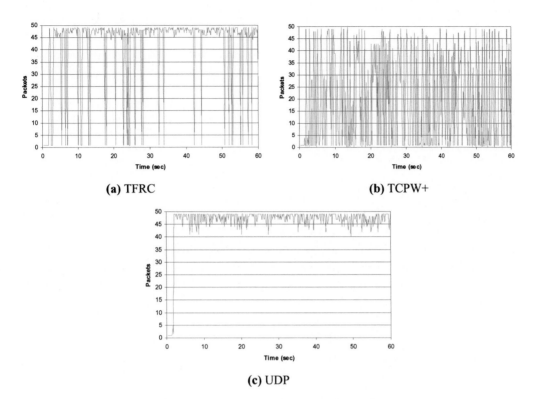

(a) TFRC

(b) TCPW+

(c) UDP

Figure 9. Queue Length of R3 (20 MPEG flows)

Table 4. Average Queue Length of R3 (20 MPEG flows)

Protocol	Average Queue Length
TFRC	41.9
TCP Westwood+	19.1
UDP	46.2

UDP achieves the highest goodput rates (30-50 flows), when the demand exceeds the bottleneck channel capacity, since the protocol steadily transmits at application rate regardless of the prevailing network conditions (Fig. 6). TFRC also exhibits efficient bandwidth utilization, outperforming TCPW+, regardless of link multiplexing. The rate-based protocol manages to control the tradeoff between responsiveness and smoothness. The equation-based responses of TFRC enable the protocol to compare very favorably with the free-transmitting UDP over a wide range of network dynamics. On the other hand, TCPW+ yields a notable inefficiency in terms of goodput performance (Fig. 6). Despite the improvements over the initial version of Westwood, TCPW+'s algorithm still does not obtain accurate estimates in heterogeneous environments, failing to achieve full utilization of the available bandwidth. Apparently, the protocol is sensitive to the disturbances caused by interfering FTP traffic. This observation is profound in the case of scarce bandwidth (high contention), where the sending rate is diminished. A comparative view at the traces of queue length (Fig. 9), as well as at Table 4 reveals that TCPW+ can hardly built up long queues at the buffer, due to its tendency to overestimate the available bandwidth. Furthermore, TCPW+ slows down the transmission in response to the diverse transient errors.

Since efficient link utilization does not necessitate an improved performance on video delivery, we also demonstrate statistics from delayed packets in order to quantify protocol efficiency. Fig. 7 illustrates that TFRC achieves the timely delivery of most packets inducing minimal impairments on perceived video quality. The protocol effectively smoothes transmission gaps and confines short-term oscillations in the sending rate. Despite the increased contention, TFRC flows are immunized from the disturbances induced by the corporate FTP connections, inline with the unresponsive UDP. However, UDP causes long and variable delays that degrade video quality (Fig. 7), due to the rapidly growing queues and buffer overflows across the bottleneck link (Fig. 9c).

In terms of fairness, Fig. 8 depicts that both TCP protocols achieve a relatively fair behavior for a wide range of contention adjustments, despite the diverse flow RTTs. On the other hand, UDP, which does not anticipate fairness, is not able to handle bandwidths sharing efficiently. In the event of high contention, competing UDP flows with aggressive mood may easily cause undesirable implications for the network.

5.3   Video Delivery vs. Error Rates

Finally, we performed a set of experiments using diverse packet error rate adjustments (PER: 0.01-0.05). We simulated 20 MPEG flows for each protocol using the dumbbell topology. We also carried out the same experiments without packet errors and used them as a reference. Our objective is to demonstrate the impact of diverse packet error rates on the performance of video delivery (Figs. 10-12).

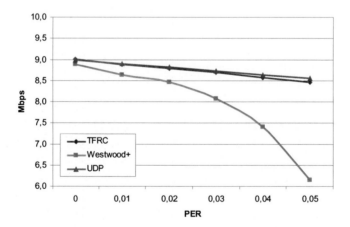

Figure 10. System Goodput (20 flows)

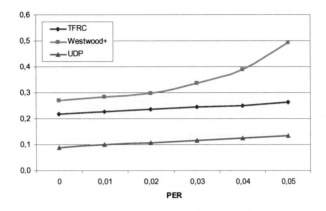

Figure 11. Packet Drop Rate (20 flows)

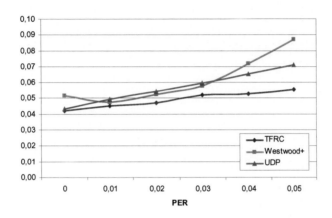

Figure 12. Delayed Packets Rate (20 flows)

TCPW+ is apparently the most sensitive protocol to the diverse bit error rates, since it does not incorporate an inherent mechanism for error detection. The specific impact is mostly reflected in bandwidth utilization (Fig. 10), where goodput performance is significantly diminished. Although the proportion of the delayed packets does not compose a major impairment factor (Fig. 12), the increasing packet drops inevitably degrade the perceived video quality and induce implications in the reconstruction of the stream at the receiver (Fig. 11). Generally, a novel strategy is needed for TCPW+, where detection and recovery should additionally account for specific characteristics of the error pattern, such as frequency and duration.

On the other hand, TFRC yields only slight sensitivity to the increasing transient errors, since its sending rate is gracefully reduced in response to the measured loss events (Fig. 10). Furthermore, the protocol maintains a satisfactory rate of delayed packets even for intensely error-prone links (Fig. 12). Similarly, UDP exhibits minor implications in the event of increasing wireless errors, as the protocol transmits at a steady rate and does not account for any type of error.

6 Conclusions

We have demonstrated the challenges and limitations of UDP and selected TCP mechanisms in terms of video delivery. In our study, we focused on TCP protocols that dynamically exploit bandwidth availability relying on precise measurements of current conditions. Despite the superiority of such mechanisms over "blind" congestion control schemes, we showed that network heterogeneity may still impact protocol performance, and consequently impair the perceived video quality. We identified that protocol efficiency over heterogeneous environments is strictly related with effective bandwidth utilization and minimal transmission gaps. In this context, we reach the outcome that TFRC is the most prominent solution among the protocol tested. Relying on equation-based responses, the protocol effectively controls the tradeoff between responsiveness and smoothness, delivering smooth video over a wide range of network and session dynamics. TFRC's robust behavior in conjunction with the inadequate performance of TCPW+, as well as other widely deployed TCP implementations [13], outline the efficiency of equation-based congestion control for time-sensitive traffic. Furthermore, we showed that UDP traffic has occasionally negative impact not only for the systemwide behavior, but also for the supporting application as well. Consequently, end-to-end congestion control does not hurt real-time traffic; inversely, it occasionally results in performance gains.

References

[1] D. G. Andersen, A. C. Snoeren and H. Balakrishnan, *"Best-Path vs. Multi-Path Overlay Routing"*, In Proc. of *Internet Measurement Conference (IMC) 2003*, Miami, Florida, USA, October 2003

[2] D. Bansal and H. Balakrishnan, "Binomial Congestion Control Algorithms", In Proc. of *IEEE INFOCOM 2001*, Anchorage, Alaska, USA, April 2001

[3] S. Blake, D. Black, M. Carlson, E. Davies, Z. Wang, and W. Weiss, "An Architecture for Differentiated Services", *RFC* **2474**, December 1998

[4] L. Cai, X. Shen, J. Pan, and J. Mark, "Performance analysis of TCP-friendly AIMD algorithms for multimedia applications", *IEEE Transactions on Multimedia*, **7**, 2, April 2005, pp. 339-355

[5] D. Chiu and R. Jain, "Analysis of the increase/decrease algorithms for congestion avoidance in computer networks", *Journal of Computer Networks*, **17**, 1, June 1989, pp. 1-14

[6] N. Feamster, D. Bansal and H. Balakrishnan, "On the Interactions Between Layered Quality Adaptation and Congestion Control for Streaming Video", In Proc. of *11th Int/nal Packet Video Workshop*, Kyongju, Korea, May 2001

[7] S. Floyd, M. Handley, J. Padhye, and J. Widmer, "Equation-Based Congestion Control for Unicast Applications", In Proc. of *ACM SIGCOMM 2000*, Stockholm, Sweden, August 2000

[8] S. Floyd and V. Jacobson, "Random Early Detection gateways for Congestion Avoidance", *IEEE/ACM Transactions on Networking*, **1**, 4, August 1993, pp. 397-413

[9] L. Grieco and S. Mascolo, "Performance evaluation and comparison of Westwood+, New Reno, and Vegas TCP congestion control", *ACM Computer Communication Review*, **34**, 2, April 2004, pp. 25-38

[10] V. Jacobson, "Congestion avoidance and control", In Proc. of *ACM SIGCOMM '88*, Stanford, USA, August 1988

[11] S. Mascolo, C. Casetti, M. Gerla, M. Sanadidi, and R. Wang, "TCP Westwood: Bandwidth Estimation for Enhanced Transport over Wireless Links", In Proc. of *ACM MobiCom '01*, Rome, Italy, July 2001

[12] P. Papadimitriou and V. Tsaoussidis, "Assessment of Internet Voice Transport with TCP", *Int/nal Journal of Communication Systems (IJCS)*, Wiley Academics, **19**, 4, May 2006, pp. 381-405

[13] P. Papadimitriou and V. Tsaoussidis, "On Transport Layer Mechanisms for Real-Time QoS", *Journal of Mobile Multimedia*, Rinton Press, **1**, 4, January 2006, pp. 342-363

[14] V. Paxson, "End-to-End Packet Dynamics", *IEEE/ACM Transactions on Networking*, **7**, 3, 1999, pp. 277-292

[15] K. Ramakrishnan and S. Floyd, "A proposal to add explicit congestion notification (ECN) to IP", *RFC* **2481**, January 1999

[16] S. Shenker, C. Partridge and R. Guerin, "Specification of Guaranteed Quality of Service", *RFC* **2212**, September 1997

[17] W. Stevens, "TCP Slow Start, Congestion Avoidance, Fast Retransmit, and Fast Recovery Algorithms", *RFC* **2001**, January 1997

[18] V. Tsaoussidis and I. Matta, "Open issues on TCP for Mobile Computing", *Journal of Wireless Communications and Mobile Computing*, Wiley Academics, **2**, 1, February 2002, pp. 3-20

[19] V. Tsaoussidis and C. Zhang, "TCP Real: Receiver-oriented congestion control", *Journal of Computer Networks*, **40**, 4, November 2002, pp. 477-497

[20] V. Tsaoussidis and C. Zhang, "The dynamics of responsiveness and smoothness in heterogeneous networks", *IEEE Journal on Selected Areas in Communications (JSAC)*, **23**, 6, June 2005, pp. 1178-1189

[21] Y. R. Yang, M. S. Kim and S. S. Lam, "Transient Behaviors of TCP-friendly Congestion Control Protocols", In Proc. of *IEEE INFOCOM 2001*, Anchorage, Alaska, USA, April 2001

[22] Y. R. Yang and S. S. Lam, "General AIMD Congestion Control", In Proc. of 8th *Int/nal Conference on Network Protocols (ICNP)*, Osaka, Japan, November 2000

[23] C. Zhang and V. Tsaoussidis, "TCP Real: Improving Real-time Capabilities of TCP over Heterogeneous Networks", In Proc. of the 11th *IEEE/ACM NOSSDAV*, Port Jefferson, New York, USA, June 2001

In: Mobile Multimedia: Communication Engineering ...   ISBN: 1-60021-207-7
Editors: I.K. Ibrahim and D. Taniar pp. 137-156   © 2006 Nova Science Publishers, Inc.

Chapter 8

UWB BLUETOOTH: STRATEGIES FOR INTEGRATION

John Nelson [*] *and Dorel Picovici* [**]

Department of Electronic and Computer Engineering,
University of Limerick, Limerick, Ireland

Abstract

Bluetooth™ is one of the most commercially successful wireless technologies with widespread adoption enabling a wide range of devices to communicate without wires. In contrast, Ultrawideband systems are emerging that promise to address the many perceived inadequacies of Bluetooth™ such as relatively high complexity, high energy consumption and low data rates. However, the Bluetooth™ success is in spite of these. It is a complete communication system that supports service discovery and a wide range of existing and new applications through the adoption of standardized profiles. From a multimedia service-provisioning viewpoint, the Bluetooth™ community learned rather quickly that there were significant system level challenges to overcome in order to allow users and devices to use the wireless medium effectively. This meant significant emphasis on characterizing and defining device capabilities, services and applications, and on enabling easy access by the end users. This chapter introduces the current Bluetooth™ system and overviews the emerging Ultrawideband technology and systems, outlining the major components of each. It presents a broad summary of the breadth of Bluetooth™ technology from the physical layer to the standardized application level profiles. For Ultrawideband, an insight is given into the standardization efforts and targeted applications. The chapter finishes by briefly looking at possible strategies for their integration.

Introduction

In 2005, the Bluetooth™ Special Interest Group (SIG)[1], responsible for the standardization and promotion of Bluetooth, announced that it would consult with the Ultra wideband (UWB) development groups to address how the technologies could be integrated to better serve the

---

[*] E-mail address: john.nelson@ul.ie
[**] E-mail address: dorel.picovici@ul.ie
[1] Two major Bluetooth web sites www.bluetooth.com *the official Bluetooth website*, and www.bluetooth.org *the official membership site*, provide extensive and definitive information on all aspects of Bluetooth.

Bluetooth™ sector. Bluetooth™ is differentiated in that it is a complete end-to-end communication system enabling devices to communicate and support a wide range of applications. It was driven initially to address specific applications such as cordless headsets and wireless serial port communications but has evolved to support a range of relatively low data rate multimedia applications. Hence, the initial Bluetooth™ development required and was constrained by relatively low transmission speed requirements.

On the other hand, the UWB promoter groups are focusing on the RF physical (PHY) and medium access (MAC) layers targeting very low interference, very high data rates and very low power consumption. Similar to Bluetooth, the emerging UWB standards are suitable for transmitting data between a wide range of devices within a short range, with support at the data link level for an extensive set of existing and future wired and wireless applications. UWB has adequate capacity to simultaneously support multiple multimedia audio and video streams, and yet transmit below the noise levels of most other systems such that it is non-interfering and even undetectable by them.

The Bluetooth™ initial target markets of cordless headsets and low data rate applications have been superseded by demands and requirements for higher data rate wireless multi-media support yet retaining low power. There is significant pressure on Bluetooth™ to improve its data rates, while at the same time the industrial driven UWB bodies need a wide application base to capture market share, in particular given recent high data rate developments within 802.11 based Wireless LAN. Bluetooth™ so far has only incrementally improved its data rates, but a UWB approach would enable an immediate and future proof solution. For Bluetooth, the issue is whether to incorporate the work of one or more of the emerging UWB radio technologies or to develop a new Bluetooth™ specific physical layer.

This chapter outlines Bluetooth™ as a system, emphasizing the aspects that have led to its success. It continues with a brief overview of UWB. Finally, a number of alternative integration strategies for Bluetooth™ and UWB are considered focusing on their impact on the whole Bluetooth™ system.

Bluetooth

Bluetooth™ is a relatively low cost, low power robust wireless technology targeted at providing short-range communication between devices using low form factor solutions. It is based upon a single specification that is globally accepted. The Bluetooth™ Special Interest Group (SIG), an industrial consortium with thousands of members, drives the specification and it is in its 4th major version since the original Version 1.0. These versions are 1.0 (Jul 1999), 1.1(Feb 2001), 1.2 (Nov 2003) and 2.0(Aug 2004). Later versions typically add new functionality and address errata from previous versions. The earlier versions are scheduled for deprecation. In general though there is backwards compatibility. This is important given that over 500 million units with Bluetooth™ capability have been shipped up until the end of 2005, half that number alone in 2005, thereby making it the most widely supported unlicensed wireless standard in its class.

Bluetooth™ operates in the unlicensed 2.4GHz band, part of the industrial, scientific and medical (ISM) band. It is robust due to a frequency hopping system (adaptive since version 1.2) where the signal hops over various channels to limit interference, and to its adoption of strong encryption and authentication. Two outdoor ranges of 10m and 100m are defined, for Class 2 and Class 1 devices, respectively.

The major success of Bluetooth™ is not just in defining the radio technology, but also largely due to innovations in defining usage profiles, providing strong compliance and interoperability insurance through the Bluetooth™ Qualification Program, and comprehensive and evolving test definitions. From the outset, the barriers to ensure widespread adoption were identified and addressed. The strongly promoted use of frequent UnPlugFests allowed early issue resolution. Unplugfests are interoperability test events run by the Bluetooth™ SIG to allow teams of development engineers to verify interoperability. Early problems with ease of use have now largely been overcome, and the major anomalies in the specification have been resolved. Detailed Bluetooth™ Implementation Conformance Statement (ICS), Protocol Implementation Conformance Statement (PICS), and Profile Implementation Conformance Statements (Profile-ICS) for all applicable protocols and profiles are adopted. Furthermore, the Bluetooth™ SIG in Aug 2005 approved a new qualification program, which was designed to simplify the process, improve interoperability, ensure consistency and reduce qualification costs.

The initial separation of Bluetooth™ into two major parts: the host and host controller with a well-defined interface, the Host to Controller Interface (HCI) between the two has ensured extensibility. Indeed, the identification of distinct product types containing Bluetooth™ wireless technology has allowed a competitive market to develop and evolve. These Bluetooth™ products types are [1]:

- Bluetooth™ End Product
- Bluetooth™ Host Subsystem Product
- Bluetooth™ Controller Subsystem Product
- Bluetooth™ Profile Subsystem Product
- Bluetooth™ Component Product
- Bluetooth™ Development Tool
- Bluetooth™ Test Equipment

This classification by product type into end products, subsystems and components, allows end products to be built from compliant parts. Furthermore, although the core specification is very large, many features are optional, allowing product evolution and differentiation.

Bluetooth™ Version 1.2 adopted a new architectural overview in terms of structure and nomenclature, with more precise language similar to the IEEE communication standards. New functionality included faster connection setup, adaptive frequency hopping, extended synchronous connection oriented links, enhanced error detection and flow control, enhanced synchronization capability, and enhanced flow specification. Also, some earlier functionality was deprecated. Bluetooth™ Version 2.0 introduced EDR (Enhanced Data Rate) supporting asymmetric data throughput up to 2.1 Mbps, resulting in overall lower power due to the reduced duty cycle.

Bluetooth™ Architecture

The Bluetooth™ physical layer (Bluetooth™ RF) uses frequency hopping over a set of up to 79 Frequency Hopping Spread Spectrum (FHSS) carriers, to provide an original gross Basic Rate of 1 Megabit per second (1Mbps), and with the Version 2.0 Enhanced Data Rates of either 2 or 3Mbps. The 1Mbps Basic Rate resulted in symmetric net data rates of 434Kbps or asymmetric paired channels of a maximum forward channel of 723Kbps with a corresponding maximum reverse channel of 57Kbps. The physical channel adopts time division duplex as shown in Figure 1 with a nominal 625us transmit and receive slot time, which are normally paired, Receive or transmit slots can be extended to 3 or 5 slots for improved data throughput.

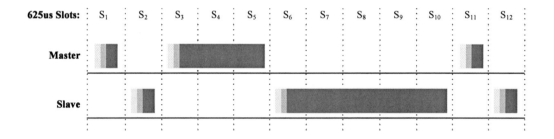

Figure 1 625us slot structure showing 1 slot, 3 slot and 5 slot packets.

The baseband packet consists of an access code, header and payload as outlined in Figure 2. The normal 72 bit access code consists largely of a synchronization part known as the syncword, followed by the 54 bit header, which contains well protected link control information. To allow for radio channel switching the maximum single slot packet is 366us resulting in a 240 bit gross data payload. The original design was largely driven by 64Kbps voice, thereby the 240 bit payload and the 800 transmit and receive paired slots per second allowed one, two or three voice connections by carrying 10, 20 or 30 8-bit samples in each payload, using the HV1, HV2 and HV3 packet types, respectively. Two types of basic rate data packets are supported, the DMx and DHx, indicating data medium and high, respectively, where x indicates the number of slots e.g. DM3 is a 3-slot medium packet type protected by both forward error correction and a cyclic redundancy check. The payloads range from maximum 17 bytes for a DM1 to 339 bytes for a DH5.

Access Code	Header	Payload
68-72 bits	54 bits	0-2744 bits

Figure 2 The physical channel basic rate packet format

A physical channel can be shared by a group of devices known as a piconet by synchronizing to a common clock of a single master device. The other devices are known as slaves. The common clock based on a 312.5us tick, and the Bluetooth™ address of the master allows the piconet to use a statistically unique pseudo random frequency hopping pattern. In Version 1.2 a subset of the 79 channels can be adopted by using the specified adaptive frequency hopping mechanism through frequency classification. Duplex transmission is supported using Time Division Duplex (TDD). A physical connection only exists between the piconet master and each slave. The physical link is used as a transport for one or more logical links that support unicast synchronous, asynchronous and isochronous traffic, and broadcast.

The complete Bluetooth™ architecture encompasses the protocols shown in figure 3. However, only a subset of the layer service access points is specified. In total, the Bluetooth™ standard defines a core specification, an additional set of supporting protocols, a set of application centric profiles, HCI transport mechanisms, and comprehensive test cases.

Basic Network Encapsulation Protocol
Transports other protocols

Object Exchange
File Transfer, Vcard

RFCOMM with TS 07.10
Serial Port Emulation

Service Discovery
Explores the capabilities of other devices

Host Controller Interface
A transport mechanism between parts of a split stack.

Baseband
The Link Controller, Radio Control and a Device Manager.

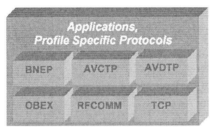

Audio/Video Control Transport Protocol
Controls AV features

Audio/Video Control Transport Protocol
Transports AV streams

Telephony Control Protocol
Call control/management (Q.931)

L2CAP Logical Links
Multiplexing, Segmentation, QoS and Connectionless data.

The Link Manager
Configuration & Security of the underlying baseband links.

Figure 3 The Bluetooth™ System Architecture

The lowest three layers form a subsystem known as the Bluetooth™ controller. The protocols and services including the L2CAP layer and above are known as the Bluetooth™ host. The important Host to Controller Interface (HCI) provides a set of defined transport mechanisms between parts of a split stack (where there are separate Bluetooth™ controller and host). The HCI supports a comprehensive standardized set of commands for data transfer and control.

The Baseband layer controls the timing and packet formations, enforces the Bluetooth™ low level timing requirements, implements Bluetooth™ Piconet and Device Clocks, frequency selection, packet type bit-stream processing etc. It contains the link controller (LC), which schedules activities associated with a Bluetooth™ link, selects the active piconet or operation to perform next e.g. inquiry, page etc. The Link Controller States are Standby, Inquiry, Inquiry Scan, Page, Page Scan, Connection-Active, Connection-Hold, Connection-Sniff, Connection-Park. The Hold, Sniff and Park operational modes support low piconet management and low power.

The link manager (LM) layer controls and negotiates all aspects of the operation of the Bluetooth™ connection between two devices by creating, modifying and releasing logical links. It takes commands usually from the Host Controller and interfaces with the device's Baseband and Link Controller, and the peer device's LM via the LMP protocol to carry out the commands. It establishes and configures data and voice links, through the asynchronous connectionless (ACL) and synchronous connection oriented (SCO) links. In addition, it enforces Low-Power modes, incorporates quality of service negotiation and radio power control. This layer is very Bluetooth™ specific and supports the following groups of procedures:

- Connection control, for connection establishment, detach, monitoring and controlling
- Security, supporting authentication and encryption
- Information requests for timing, clock offset, version, features, and name
- Role Switch to allow master and slave role changes by deriving slot offset, and invoking the switch request
- Modes of operation for low power such as Hold, Sniff, and Park
- Logical transports supporting SCO/eSCO links request, removal and change, which are typically used for voice.
- Test mode to activate and control the defined certification/compliance test procedures

Each group may contain multiple procedures; the connection control group includes detailed procedures for Connection Establishment, Connection Detach, Power Control, Adaptive Frequency Hopping, Channel Classification, Link supervision, Channel quality driven data rate change, Quality of Service, Paging Scheme parameters and Control of multi-slot.

The HCI is a well-defined interface mainly to allow the host comprehensive access to the Bluetooth™ controller, and for the host controller to report state changes and to deliver data to the host. The host accesses the controller typically to send control commands, send and receive data and voice packets. The host receives events from the controller when a command has been processed or something happens. There are various HCI transport layers transparently supported [2] such as UART, Secure Digital, USB, and 3-wire UART. A summary of the command and event groups is presented in Table 1 to indicate the scope of the interface. The majority of the commands are very closely coupled to the underlying host controller functionality, and therefore they are not generically reusable for alternative wireless technologies.

Table 1 HCI Command and Event Groups

Group	Brief description
Device Setup	The device setup commands are used to place the Controller into a known state.
Controller Flow Control	The controller flow control commands and events are used to control data flow from the host to the controller.
Controller Information	The controller information commands allow the host to discover local information about the device.
Controller Configuration	The controller configuration commands and events allow the global configuration parameters to be configured.
Device Discovery	The device discovery commands and events allow a device to discover other devices in the surrounding area.
Connection Setup	The connection setup commands and events allow a device to make a connection to another device.
Remote Information	The remote information commands and events allow information about a remote device's configuration to be discovered.
Synchronous Connections	The synchronous connection commands and events allow synchronous connections to be created and managed.
Connection State	The connection state commands and events allow the configuration of a link, especially for low power operation.
Piconet Structure	The piconet structure commands and events allow the discovery and reconfiguration of a piconet.
Quality of Service	The quality of service commands and events allow quality of service parameters to be specified.
Physical Links	The physical link commands and events allow the configuration of a physical link.
Host Flow Control	The Host flow control commands and events allow flow control to be used towards the host.
Link Information	The link information commands and events allow information about a link to be read.
Authentication and Encryption	The authentication and encryption commands and events allow authentication of a remote device and encryption of the data link(s).
Generic Events	The generic events can occur due to multiple commands, and covers events that can occur at any time.
Testing	The testing commands and events allow a device to be placed into test mode.

The core Host layers are the, SDP, RFCOMM and L2CAP [1]. SDP, the Service Discovery Protocol allows devices to publish and discover in a standardized manner, the services which are supported on a particular Bluetooth™ device. RFCOMM, is a serial cable emulation protocol, based on the TS07.10 protocol, which provides the necessary functionality to emulate serial ports e.g. COM ports, allowing seamless migration of the many legacy applications to Bluetooth. It supports multiple simultaneous ports using multiplexing and flow control.

The L2CAP (Logical Link Control and Adaptation Protocol) layer and protocol is a data link layer equivalent protocol that provides connection-oriented and connectionless data services to its user protocols supporting packet lengths of up to 64K bytes. It implements segmentation and reassembly, multiplexing, and group abstractions. Bluetooth™ Version 1.2 added enhanced error protection and flow control for L2CAP, before that L2CAP relied on the expected reliability of the baseband link protocol. However, the protection on the high data rate packets in particular was shown to be inadequate on poor channels, thereby requiring the higher-level protection.

The Bluetooth™ Core System Architecture [1] is defined as covering the Bluetooth™ controller, the HCI interface, the service discovery protocol (SDP) and the overall profile requirements as specified in the Generic Access Profile (GAP), and is outlined in figure 4. Bluetooth™ relies strongly on the definition of the inter device communication protocols to ensure interoperability, thereby specifying the radio, the Link Control protocol, the Link Manager protocol (LMP) and the L2CAP protocol. As a result, there is significant opportunity for implementation differentiation of the actual protocol stacks. In addition, the core architecture identifies two L2CAP blocks: the channel manager and the L2CAP resource manager; two Baseband Layer blocks: the Link Controller and Baseband Resource Manager; and finally an LMP/Baseband cross layer device manager.

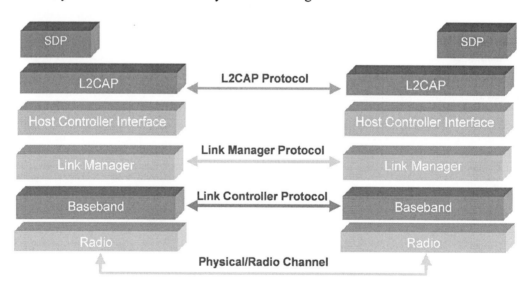

Figure 4 The core layers and inter device protocols.

The L2CAP protocol allows multiple applications to share the data link using managed channels between two connected devices or groups of devices. Channels are managed on each device by channel endpoint identifiers, and can have defined quality of service (QoS), and per channel flow control. The LMP is used to communicate between two devices connected by an ACL logical transport (i.e. between the master and each slave). This includes the set-up and control of logical transports and logical links, and for control of physical links. The protocol consists of a set of LMP messages, also known as LMP PDUs (Protocol Data Units). The LMP Message payload is relatively simple consisting of a single bit Transaction Identifier (to indicate whether the transaction was initiated by Master or Slave), a 7 or 15 bit Operation Code (Opcode), and the message parameters. All messages are normally transferred as DM1 packets on the reliable ACL. In V1.2 the LMP protocol was modified for future extensibility. Finally, the Link Controller protocol is carried in the packet header and controls acknowledgement, sequencing, and flow control of packets.

Applications and Multimedia

The Bluetooth™ SIG and the participant organizations have addressed a very broad range of applications including multimedia, targeting a wide set of devices from consumer devices, to cellular handsets, and PC accessories. The main focus has been on providing very high quality voice, including stereo headsets, and streaming data services through the standardization of the associated profiles. Even so, the applications have often been constrained by the available low data rates.

The Bluetooth™ Profiles, summarized in New Bluetooth™ specific protocols were developed to support the profiles. Table 3 identifies all the host level protocols. In addition to L2CAP and SDP defined in the core specification, six other protocols AVCTP[3], AVDTP[4], BNEP[5], OBEX[6], RFCOMM[7] and TCP[8] are separately specified. The Audio/Video Control Transport Protocol (AVCTP), is used to transport command and response messages for controlling Audio Video features, and uses point to point signaling over connection oriented L2CAP channels. The Audio/Video Distribution Transport Protocol specifies the connections and streaming for audio or video over Bluetooth via L2CAP, and is based on the Real-Time Protocol (RTP) and RTP Control Protocol (RTCP). The Bluetooth Networking Encapsulation Protocol supports the carrying of various existing network protocols such as IPv4, IPv6, and IPX over L2CAP. The Obex protocol enables the exchange of data objects such as vCards representing electronic business cards. The RFCOMM protocol supports the transport of up to 60 simultaneous serial port connections, with provision to emulate the 9 circuits of an RS-232 serial port. The TCP protocol supports telephony call control based on the ITU Q.931.

Table 2, have been a key contributor to Bluetooth's success especially in ensuring interoperability between various manufacturers' implementations. The profiles each have a separate specification document except for the Generic Access Profile[1], which is a part of the core specification. Profiles typically describe how a subset of the Bluetooth™ core functionality can be adopted to support one or more user scenarios, by defining a selection of the mandatory or optional messages, procedures and parameter value(s). Each profile generally can be thought of as a vertical cut through the core layers, incorporating additional functionality and/or protocols as required. Significant effort by the Bluetooth™ SIG is focused on defining profiles in addition to developing the core specification. Currently, there are 25 separate profile specifications available, increasing from 13 in Version 1.1.

New Bluetooth™ specific protocols were developed to support the profiles. Table 3 identifies all the host level protocols. In addition to L2CAP and SDP defined in the core specification, six other protocols AVCTP[3], AVDTP[4], BNEP[5], OBEX[6], RFCOMM[7] and TCP[8] are separately specified. The Audio/Video Control Transport Protocol (AVCTP), is used to transport command and response messages for controlling Audio Video features, and uses point to point signaling over connection oriented L2CAP channels. The Audio/Video Distribution Transport Protocol specifies the connections and streaming for audio or video over Bluetooth via L2CAP, and is based on the Real-Time Protocol (RTP) and RTP Control Protocol (RTCP). The Bluetooth Networking Encapsulation Protocol supports the carrying of various existing network protocols such as IPv4, IPv6, and IPX over L2CAP. The Obex protocol enables the exchange of data objects such as vCards representing electronic business cards. The RFCOMM protocol supports the transport of up to 60 simultaneous serial port connections, with provision to emulate the 9 circuits of an RS-232 serial port. The TCP protocol supports telephony call control based on the ITU Q.931.

Table 2 Bluetooth™ Protocol Specifications

Profile Abbreviation	Profile Title	Publish Date
A2DP	Advanced Audio Distribution Profile	04-Jun-2003
AVRCP	Audio/Video Remote Control Profile	04-Jun-2003
BIP	Basic Imaging Profile	25-Jul-2003
BPP	Basic Printing Profile	10-Feb-2004
CIP	Common ISDN Access Profile	16-Nov-2002
CTP	Cordless Telephony Profile	22-Feb-2001
DUN	Dial-Up Networking Profile	22-Feb-2001
FAX	Fax Profile	22-Feb-2001
FTP	File Transfer Profile	22-Feb-2001
GAP	Generic Access Profile	*In Core Spec*
GAVDP	Generic Audio/Video Distribution Profile	24-May-2003
GOEP	Generic Object Exchange Profile	22-Feb-2001
HCRP	Hardcopy Cable Replacement Profile	23-Sep-2002
HFP 1.5	Hands-Free Profile 1.5	25-Nov-2005
HFP	Hands-Free Profile	29-Apr-2003
HSP	Headset Profile	22-Feb-2001
HID	Human Interface Device Profile	24-May-2003
ICP	Intercom Profile	22-Feb-2001
OPP	Object Push Profile	22-Feb-2001
PAN	Personal Area Networking Profile	20-Feb-2003
SPP	Serial Port Profile	22-Feb-2001
SDAP	Service Discovery Application Profile	22-Feb-2001
SAP	SIM Access Profile	12-May-2005
SYNCH	Synchronization Profile	22-Feb-2001
VDP	Video Distribution Profile	08-Sep-2004
WAPB	WAP Bearer	22-Feb-2001

The most relevant profiles for multimedia are the A2DP [9], AVRCP [10], GAVDP [11] and VDP [12]. VDP defines support for video streams, with mandatory support for the H.263 baseline, and optional support for the MPEG-4 visual simple profile and the H.263 profiles 3 and 8. There are significant dependencies, VDP is dependent on GAP and GAVDP; it has dependencies on the Host Bluetooth™ protocols, such as SDP, L2CAP, and AVDTP; and it specifies requirements on the controller protocols. This means that the profiles are dependent and closely coupled to the core architecture, e.g. specifying for the Link Controller the packets that shall be supported.

Table 3 Bluetooth™ Host Protocol Specifications

Protocol Abbreviation	Protocol Title	Publish Date
AVCTP	Audio/Video Control Transport Protocol	11-Jun-2003
AVDTP	Audio/Video Distribution Transport Protocol	11-Jun-2003
BNEP	Bluetooth Network Encapsulation Protocol	14-Mar-2003
L2CAP	Logical Link Control and Adaptation Protocol	*In Core Spec.*
OBEX	Object Exchange (OBEX)	22-Feb-2001
SDP	Service Discovery Protocol	*In Core Spec*
TCP	Telephony Control Protocol (TCP)	22-Feb-2001
RFCOMM	RFCOMM with TS 07.10	05-Jun-2003

Ultra-wideband (UWB)

Ultra-wideband (UWB) communication techniques have witnessed significant increased attention in both academia and industry for applications in short-range wireless communications. This is due to the appealing features such as high data rate, less path loss, better immunity to multipath propagation, availability of low-cost transceivers, low transmit power, low interference, flexibility, robustness and so on. These features made UWB an excellent candidate for a variety of wireless applications. On the other hand research and development efforts are continuously being made to overcome the technical challenges in UWB deployment, such as UWB channel characterization, transceiver design, antenna design for, and synchronization of, very short pulses, performance degradation due to multiple access interference and narrowband jamming, employment of higher order modulation schemes to improve capacity or throughput, coexistence and interworking with other narrowband wireless systems and design of link and network layers to benefit from UWB transmission characteristics.

A UWB system is generally defined as any radio system that has a 10-dB bandwidth larger than 20% of its central frequency, or a 10-dB bandwidth equal to or larger than 1.5 GHz if the center is higher than 6 GHz [13]. Although, in several publications UWB is considered a recent breakthrough in wireless communications, in reality its technology has experienced many years of developments [14]. The concept of UWB communication originated with Marconi and Hertz in the late 1890s, when spark gap transmitters were used to induce pulsed signals with a very wide bandwidth. Spark transmissions created broadband interference and did not allow coordinated spectrum sharing. Due to this inconvenience, the wideband communication was abandoned in favor or narrowband or tuned, radio transmitters. However, in mid-1980s the Federal Communications Commission (FCC) encouraged an entirely new type of wideband communications when it allocated the Industrial Scientific and Medical (ISM) bands for unlicensed spread spectrum and wideband communication use. In 2002 the FCC approved the deployment of UWB spectrum in the 3.1-10.6 GHz band [15]. This innovative spectrum allocation is considered to be responsible for the remarkable growth in Wireless Local Area Networks (WLAN) and wireless fidelity (Wi-Fi) today, as it led the communications industry to investigate the merits and implications of wider bandwidth communications than had been previously used for consumer applications. The interest in UWB increased exponentially since the FCC allowed its commercial deployment with a given spectral-mask requirement for both indoor and outdoor applications.

UWB Standards

To understand the UWB standardization activities warrants a brief look at the IEEE based standardization for wireless systems. IEEE 802.15 is an international standards working group for the Wireless Personal Area Networks (WPANs). The group is responsible for creating a variety of WPANs standards and as shown in **Figure 5** is divided in four major task groups.

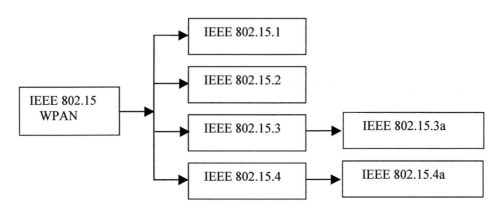

Figure 5 Organization of the IEEE 802.15

The IEEE 802.15.1 standard includes the following:

- It is based on Bluetooth v1.1 standard
- Data traffic can reach a maximum of 732 kbps (unidirectional)
- Radio link is up to 10 m (short-range)
- Includes an adaptation of the Bluetooth Media Access Control (MAC) and physical (PHY) layers as well as Logical Link Control/MAC (LLC/MAC) interface.
- Approved by IEEE Standard Association (IEEE-SA) on April 15, 2002 and was published on June 14, 2002.

IEEE 802.15.2 is concerned with the technical issues that appear when two wireless systems share the same environment of operation [15]. For these cases the task group has two goals:

- The first goal quantifies the interference effects between WLAN and WPAN wireless devices.
- The second goal is to establish mechanisms for coexistence of WLAN (IEEE 802.11) and WPAN (IEEE 802.15.1) at both MAC and PHY layer. The receiver sensitivity degradation and the reduction of throughput in the presence of an interferer represent typical metrics for evaluating the coexistence.

The IEEE 802.15.3 task group is developing WPANs with data traffic up to 55Mbps. The standard uses five 15 MHz channels in the 2.4 GHz ISM band, two of which interfere with IEEE 802.11b traffic. In order to achieve five different data rates (11, 22, 33, 44, and 55 Mbps) three modulation schemes (QPSK, DQPSQ, 16/32/64-QAM) and coding (trellis-coded modulation) are used. The MAC layer described by the IEEE 802.15.3 allows for multimedia quality of service (QoS), power management, and ad-hoc networking support.

The UWB focused IEEE 802.15.3a task group (also known as "TG3a") was formed in late 2001 and disbanded in January 2006. The main task of this group is to establish technical requirements for the development of high-speed and low-power WPAN physical layer. As shown in Table 4 these requirements will guarantee that the high data rate physical layer drafted by TG3a can be easily included into WPAN devices, which have MAC and network layers already implemented in CMOS technology.

Table 4 Summary of the technical requirements for IEEE 802.15.3a [16]

Parameter	Value
Antenna practicality	Size and form factor consistent with original device
Cost	Similar to Bluetooth
Co-located piconets	Four
Co-existence capability	Reduced interference to IEEE systems with a interfering average power at least 6 dB below the minimum sensitivity level of non-802.15.3a devices.
Data rates	110, 220 and 480 Mbps (optional)
Interference susceptibility	Robust to IEEE systems. PER <8% for a 1024 byte packet
Location awareness	Location information to be propagated to a suitable management entity.
Power consumption	100 mW and 250 mW.
Power management modes	Power save and wake up capabilities
Range	10 m, 4 m and below
Scalability	Backwards compatibility with 802.15, adaptable to various regulatory regions (US, European countries and Japan)
Signal Acquisition	Less then 20 µs for acquisition from the beginning of the header

The TG3a's most commendable achievement was the consolidation of several UWB PHY specifications into two proposals:

- MultiBand Orthogonal Frequency Division Multiplexing (MB-OFDM) UWB, supported by the WiMedia Alliance, and
- Direct Sequence-UWB (DS-UWB), supported by the UWB Forum.

The IEEE 802.15.4 task group is focused on low data rate and low power WPAN (LP-WPAN). It also investigates low data rate WPAN solutions with a battery life ranging from months to several years and a very low complexity. The main features of IEEE 802.15.4 are as follows:

- Data rates of 250 Kbps, 40 Kbps, and 20 Kbps.
- Two addressing modes; 16-bit short and 64-bit IEEE addressing.
- Support for critical latency devices, such as joysticks.
- CSMA-CA channel access.
- Automatic network establishment by the coordinator.
- Fully handshaked protocol for transfer reliability.
- Power management to ensure low power consumption.
- 16 channels in the 2.4GHz ISM band, 10 channels in the 915MHz and one channel in the 868MHz band.

The proposed IEEE 802.15.4 draft was approved by IEEE-SA on May 12, 2003. The IEEE 802.15.4a was formed in March 2004 and the main interests of the group are: high precision (1 meter accuracy and better), high aggregate throughput and ultra low power.

Applications

In general UWB is considered to have potential for applications that are not fulfilled by currently available wireless short-range technologies [17]. Examples of these applications are:

1) Communications:
   - Wireless personal area networks (PANs), local area networks (LANs)
   - High speed, multi user wireless networks
   - Indoor video/data/voice distribution (digital TV sets, VCRs, audio CD/DVD, MP3 players, etc)
   - Wireless intercom system/UWB radio transceiver
   - Military applications
2) Vehicular radar systems:
   - Underground, through-wall and ocean imaging
   - Security systems used for alarming and tracking movement
   - Vehicular radar systems: airbags developments, suspensions/braking based on road conditions, vehicular entertainment and navigation.
   - Precision measurement devices

- Aviation/obstacle avoidance radar
- Military applications such as intrusion detection radars, aerial vehicles

3) Sensor networks:

- Avalanche monitoring and pollution tracking
- Precision Asset Location System
- In-building tracking and aviation ground tracking
- UWB geolocation system – for Emergency 911, fire and rescue
- Military applications – personal security and logistics

From a UWB Bluetooth perspective the focus will largely be on enabling a combined UWB and Bluetooth to support the communications applications above, as well as Bluetooth's legacy applications.

UWB Systems Design

When designing a UWB system the following models and techniques have to be taken into account: Signal models, Modulation techniques, and Demodulation techniques. Based on the processing domain the designs can be categorized into two different groups: time domain and frequency domain designs.

Time Domain Processing

The time domain processing UWB system designs have been widely presented in the literature hence the following paragraphs give a brief description of these systems.

In time domain, the UWB signals are formed from short duration pulses, generally in the order of nanoseconds. These pulses are normally created using a single basic pulse shape. This pulse train is referred to as impulse radio and has a very low duty of cycle. The data bits to be transmitted are modulated using these short radio pulses with repetition duration bigger than the time delay spread of the multipath channel. On the other side at the receiver, the resolvable multipath components are combined such that the information can be recovered with a minimum probability of error.

1) Time domain signal model

One of the most popular time domain UWB signal model is based on time hopping (TH) impulse radio (IR). From a kth user perspective, the transmitted signal of TH-IR system can be described using the following formula [18]:

$$s_{tr}^{(k)}(t^{(k)}) = \sum_{j=-\infty}^{\infty} w_{tr}(t^{(k)} - jT_f - c_j^{(k)}T_c - \delta d_{\lfloor j/N_s \rfloor}^{(k)})$$ (1)

where $t^{(k)}$ represents the kth user's clock time, T_f is the pulse repetition time, $w_{tr}(t)$ is the transmitted pulse, $c_j^{(k)}$ is the time-hopping sequence, T_c is the time difference between hops and N_s represents the number of monocycles modulated by each information bit. The data sequence $d^{(k)}_{\lfloor j/N_s \rfloor}$ is modulated with the modulation index δ. More details of this time domain signal model are given in [18].

2) Time domain modulation techniques

The time domain signal model described by equation (1) can be modulated using various time domain modulation techniques [17]. As one of the main constraints of the power spectral density (PSD) of the transmitted signal is the transmit power, the choice of time domain modulation techniques have to take into consideration both power efficiency and the effect of the modulation technique on the PSD. The most used techniques are briefly described below.

- Pulse Position Modulation (PPM). The PPM is one of the most widely used modulation techniques. With this technique the information is carried in the fine time-shift of the pulse. In order to reduce the total transmit power, the PPM techniques reduce the spectral lines by using a time-hopping sequence design [18].
- Binary Phase Shift Keying (BPSK). The BPSK is considered to be extremely power efficient. However its main drawback is that requires two polarities of pulses to be generated. This leads to a complex transceiver in comparison with the PPM case which requires only single polarity.
- Pulse Amplitude Modulation (PAM). With PAM TH-IR system the information is carried in the amplitude of the radio pulses. This technique is considered to have very poor energy efficiency [19].
- On/Off Keying (OOK). OOK represents a special case of PAM. One of the advantages over PAM is considered to be its implementation, but it also inherits PAM's poor energy efficiency [20].

3) Time domain demodulation techniques

The correlation receiver and rake receiver represents the two most widely used demodulators in UWB systems. The following paragraphs introduce these two time domain demodulation techniques.

- UWB Correlation Receiver (CR). The CR is optimised for a single bit of a binary modulated TH-IR radio signal. The received signal by the CR is correlated with the expected receive pulse (reference or template signal). Based on the sign of correlation values a binary decision is made [18]. The CR has a low cost of implementation. However, it does not take into consideration the multipath environment and it does not provide optimal detection in a multiple access system with knowledge of the time-hopping sequence.
- UWB Rake Receiver (RR). A standard RR is structured using correlators and linear combiners. The received signal by the RR is correlated with the equally delayed versions of the reference (template) radio pulse. The received signal is also sampled and multiplied by a number of tap weights before a linear combination. The number of tap weights is given by the total number of resolvable signal components. The

number of paths to be combined and the combining methods are the two major considerations when designing a UWB RR. The path selection includes the following techniques: maximum selection, partial selection and threshold selection. On the other hand the combining methods are represented by: maximum ratio combining and minimum mean square error combining. Both path selection and combining methods are detailed in [21 22 23 24].

Frequency Domain Processing

The UWB system designs in the frequency domain are based on orthogonal frequency-division multiplexing (OFDM). In an OFDM system, the channel bandwidth is equally subdivided into a number of subchannels, each channel having the bandwidth sufficiently narrow for ideal channel response approximation [17]. Having the subcarriers associated to orthogonal subchannels, the OFDM allows different data rate transmission on each subcarrier. For the UWB system in the presence of narrowband or multipath interference, the traditional OFDM has been proved to have unacceptable performance [25].

Emerging UWB candidates

Several proposed systems have been investigated by the IEEE 802.15.3a (TG3a). However, in January 2006, TG3a withdrew the project authorization, which had initiated the development of UWB standards leaving two competing proposals, which are currently the main opportunities for Bluetooth™ adoption. The first of the consequent two emerging proposals adopts MultiBand Orthogonal Frequency Division Multiplexing (MB-OFDM) UWB, supported by the WiMedia Alliance. The WiMedia Alliance[2] led by Intel, Texas Instrument and other companies is a not-for-profit association that promotes and enables the adoption, regulation, standardization and multi-vendor interoperability of ultra-wideband (UWB) worldwide. WiMedia UWB focuses on wireless personal-area networks delivering high-speed (480Mbps and beyond), low-power multimedia capabilities for the PC, CE, mobile and automotive market segments. The second proposal refers to direct sequence-UWB (DS-UWB), supported by the UWB Forum. The UWB Forum[3], an industry organization led by Motorala and Freescale is similarly dedicated to ensuring the interoperability between UWB products from multiple vendors. Likewise, the product range is considered to be mobile phones, set-top boxes, computers, televisions, digital camcorders and so on. Full details of the PHY specifications about MB-OFDM and DS-UWB are given in [26, 27]. The PHY and MAC layer of the two proposed standards are different. The MB-OFDM also referred to as MBOA, is defining a new MAC, which is still under development [28]. On the other hand, the DS-UWB standard is based on the IEEE 802.15.3 as the MAC support, which was initially designed for WPANs [29].

Strategies for Integration

In effect, four distinct approaches for UWB Bluetooth™ integration will be considered. These strategies are independent of which In all cases, existing and new high bandwidth profiles

---

[2] More information about WiMedia UWB can be found at www.wimedia.org
[3] More information about UWB Forum can be found at www.uwbforum.org

will need be adapted, developed and supported although many of higher level protocols will be reusable.

1. Bluetooth™ UDR

Bluetooth™ UWB Data Rate (UDR) is adopted for the Data Part similar to Bluetooth™ EDR introduced in Version 2.0 with a specific UWB Data Rate. In this scenario the major Bluetooth protocols are retained and the major changes are to the physical channel. The new UWB features are identified by the exchange of the LMP_features_req/res messages indicating whether the devices support the UWB extensions. New packet types will be introduced similar to the 2-DHx, 3-DHx, 2-EVx, and 3-EVx packets for EDR. This is the most straightforward to implement, improving significantly the data rates of Bluetooth, but is restricted by the legacy access codes and headers. The power saving is limited, and will show only incremental improvement over Bluetooth. Two radios and baseband are required.

2. Bluetooth™ over UWB

In this mode Bluetooth™ adopts UWB as a complete new physical layer. This will lead to significant compatibility issues but allows the best leveraging of the UWB technology, including very low power. Bluetooth is estimated to consume 50x as much power to transfer a bit as UWB. Many of the higher level profiles and protocols can be retained especially above the HCI. However, the major controller protocols will largely be redundant, including the LMP, and the HCI will need significant redefinition, although it would be possible to evolve the HCI for compatibility. Given that Bluetooth is largely used for point to point communication, the UWB capabilities could initially be evolved. By adopting UWB as the physical layer, the complete MAC functionality can be transparently leveraged e.g for contention based channel access, distributed reservation, dynamic channel selection, security, range measurement etc.

3. Bluetooth™ and UWB Coexistence (separated)

In this mode Bluetooth and UWB coexist on the same device through appropriate integration. Devices will support both Bluetooth and UWB, using separate protocol stacks and radios, allowing simultaneous usage without interference. Bluetooth can be used for device and service discovery, and backwards compatibility while UWB in parallel provides the high bandwidth data transport channels necessary to support the high data rate multimedia between compliant devices. New Service Access Points will be required to ensure synchronization, and transparency for the applications. This strategy has the advantage that the two relatively complex systems could evolve rather independently.

4. Bluetooth™ and UWB Coexistence (combined)

In this strategy, the integration is similar to the previous one, but there is much more aggressive integration of the Bluetooth controller and the UWB MAC/PHY components. If required low level scheduling and the low power modes of Bluetooth, in particular Sniff can be used to schedule non-overlapping UWB/Bluetooth access.

Conclusion

This chapter reviewed the latest Bluetooth™ V2.0 standard and identified its components, focusing on the breadth of the Bluetooth™ communication system. UWB was overviewed briefly outlining the system characteristics, and identifying the two most promising industrial driven approaches. A number of options for combining Bluetooth™ and UWB were introduced and the impact on Bluetooth™ reviewed.

The Bluetooth™ SIG for its long term survival needs to develop or attain a much higher rate physical layer (PHY), and one of the potential approaches is to adopt an Ultrawideband (UWB) based approach. There has been significant effort in developing a world wide UWB based standard by the IEEE task group, 802.15 TG3a, but this task group has recently been disbanded and two associated competing industrial groups have emerged, the WiMedia Alliance and the UWB Forum, each defining distinct UWB MAC and PHY standards with very short-range speeds of up to 480Mbps. It is to be expected that only one will be commercially successful but without proven deployments, it is too early to call.

Very recently, both groups have shown the feasibility of being integrated with a commercial Bluetooth stack, thereby demonstrating early prototypes of combined UWB Bluetooth™ systems. For Bluetooth, it is expected that the most successful market strategy will be to adopt one[4] or both of these standards and to reuse and extend the existing Bluetooth™ protocol stack and profiles ensuring that next generation Bluetooth™ devices remain backwards compatible. From the UWB Forum and WiMedia Alliance perspectives, it is arguable that Bluetooth™ will need to be one of their early applications given its large installed base.

References

[1] *Bluetooth™ SIG, Bluetooth™ Specification Version 2.0 + EDR*, Volume 1, Architecture and Terminology Overview, Nov 2004.

[2] *Bluetooth™ SIG, Bluetooth™ Specification Version 2.0 + EDR*, Volume 4, Host Controller Interface [Transport Layer], Jan 2006.

[3] *Bluetooth™ SIG, Irda Interoperability - OBEX*, (or Part F:2 Core Specification V1.1), Feb 2001

[4] *Bluetooth™ SIG, Telephony Control Protocol Specification - TCS Binary* (or Part F:3 Core Specification V1.1).), Feb 2001

[5] *Bluetooth™ SIG, RFCOMM with TS 07.10 - Serial Port Emulation*, Jun 2003

[6] *Bluetooth™ SIG, Audio/Video Control Transport Protocol*, Jun-2003

[7] *Bluetooth™ SIG, Audio/Video Distribution Transport Protocol*, Jun-2003

[8] *Bluetooth™ SIG, Bluetooth Network Encapsulation Protocol*, Mar-2003

[9] *Bluetooth™ SIG, Advanced Audio Distribution Profile*, May 2003

[10] *Bluetooth™ SIG, Audio/Video Remote Control Profile*, May 2003

[11] *Bluetooth™ SIG, Generic Audio/Video Distribution Profile*, May 2003

[12] *Bluetooth™ SIG, Video Distribution Profile*, Sept 2004.

---

[4] The Bluetooth SIG have just announced (Mar 2006) its selection of the WiMedia Alliance MB-OFDM version of ultra-wideband (UWB) for integration with current Bluetooth wireless technology.

[13] *First Report and Order in the Matter of Revision of Part 15 of the Commission's Rules Regarding UWB Transmission Systems, ET Docket 98-153*, Federal Communications Commission, FCC 02-48, Apr. 22, (2002).

[14] X. Shen, M.Guizani, R. C. Qiu and T. Lee-Ngoc, *Ultra-wideband wireless Communications and Networks,* John Wiley & Sons Ltd, UK, (2006).

[15] Revision of Part 15 of the Commission's Rules Regarding Ultra-Wideband Transmission Systems, ET Docket 98-153, Federal Communications Commission, (FCC), Apr. 22, (2002).

[16] K. Mandke, H. Nam, L. Yerramneni, C. Zuniga, and T. S. Rappaport, The Evolution of Ultra Wide Band Radio for Wireless Personal Area Networks, *High Frequency Electronics*, **2**, 5, (2003).

[17] W. Siriwongpairat, Ultra Wideband Indoor Wireless Communications, University of Maryland, June 2003.

[18] M. Z. Win and R. A. Scholtz, Impulse Radio: How it works, *IEEE Communications Letters*, **2**, 2, (1998).

[19] M. L. Welborn, System Configurations for Ultra-Wideband Wireless Networks, *IEEE Radio and Wireless Conference 2001*, RAWCON 2001,pp.5-8.

[20] J. Reed, M. R. Buehrer and D. McKinstry, Introduction to UWB: Impulse Radio for Radar and Wireless Communications, http://www.mprg.org/people/buehrer/ultra/UWB %20 tutorial.pdf.

[21] I. Bergel, E. Fishler and H. Messer, Narrowband interference suppression in time-hopping impulse-radio systems, *IEEE Conference on Ultra Wideband Systems and Technologies*, 2002, pp. 303-307.

[22] D. Cassioli, M. Z. Win, F. Vatalaro and A. F. Molisch, Performance of low complexity RAKE reception in a realistic UWB channel, *IEEE International Conference on Communications*, ICC 2002, Vol.2, pp. 763-764.

[23] J. G. Proakis, *Digital Communications*, McGraw-Hill, Fourth edition, (2001).

[24] Q. Li and L. A. Rusch, Hybrid RAKE/Multiuser Receiver for UWB, http://www.intel.com/technology/comms/uwb/download/HybridRAKEMultiuserReceive rsforUWB.pdf.

[25] D. Gerakoulis and P.Salmi, Link performance of an ultra wide bandwidth wireless in-home network, Seventh International Symposium on Computers and Communications, ISCC Proceedings, 2002, pp. 699-704.

[26] DS-UWB Physical Layer Submission to 802.15 Task Group 3a, *IEEE P802.15-04/0137rl*, March 2004.

[27] Multi-band OFDM Physical Layer Proposal for IEEE 802.15 Task Group 3a, *IEEE P.802.15-03/268r4*, March 2004.

[28] *MBOA Wireless MAC Specification for High Rate Wireless Personal Area Networks* (WPANs), Version 0.65, April 2004.

[29] *IEEE Standard for Information Technology-Telecommunications and Information Exchange between Systems-Local and Metropolitan Area Networks-Specific Requirement-Part 15.3: Wireless Medium Access Control (MAC) and Physical Layer (PHY) Specifications for High Rate Wireless Personal Area Networks* (WPANs), September 2003.

In: Mobile Multimedia: Communication Engineering ...   ISBN: 1-60021-207-7
Editors: I.K. Ibrahim and D. Taniar pp. 157-195   © 2006 Nova Science Publishers, Inc.

Chapter 9

MOBILE MULTIMEDIA: REPRESENTATION, INDEXING, AND RETRIEVAL

Bo Yang[*] and Ali R. Hurson[**]
The Department of Computer Science & Engineering
The Pennsylvania State University, University Park, PA 16802

Abstract

In the past decade, mobile multimedia has experienced an explosive development because of the technological advances in wireless communications, mobile computing, and multimedia streaming. With the proliferation of wireless networks, it is expected that multimedia data transmission and manipulation will achieve even larger growth in the next decade. At the same time, due to the characteristics of multimedia data (e.g. large data volume and lack of accurate content representation techniques), the existing wireless network protocols cannot guarantee accuracy, efficiency, and robustness when performing content-based multimedia retrieval. Hence, it is highly necessary to investigate the state-of-the-art and future trends in multimedia storage, representation, and accessing. This article is intended to discuss and analyze these issues in details. Moreover, it addresses a novel solution to the existing problems.

Content representation is one of the fundamental issues in multimedia data manipulation. As witnessed in the literature, a great deal of research has been done on this issue to facilitate content-based retrieval of multimedia data. The present content-representation models can be classified into four categories: clustering, representative-region, annotation, and decision tree. These models are introduced and then their strengths and weaknesses are compared and contrasted against each other. It will be shown that each model is suitable for a certain set of application domains and hence, there is not an efficient model that can provide a general-purpose solution to every mobile multimedia application.

Accurate and efficient content-based retrieval of multimedia data is one of the main objectives of mobile multimedia data management systems. Various content-based indexing models have been proposed on the basis of multi-dimensional feature space. However, due to the reliance on low-level feature representation, there is always a tradeoff between accuracy and efficiency. Moreover, the overall performance is greatly influenced by the distribution of the multimedia data, specially, in a heterogeneous distributed environment.

---

[*] E-mail address: byang@cse.psu.edu
[**] E-mail address: hurson@cse.psu.edu

Some recent research on multimedia representation has focused on the extraction of high-level semantics instead of low-level features. Many advantages can be obtained through the analysis of semantics: easier content representation, less space requirement, more accurate retrieval, and more scalability. However, the existing semantic-analysis models that have advanced in the literature are still in their infancy and hence, experience several deficiencies.

As an attempt to solve the problems in multimedia representation and retrieval, a novel semantic-based scheme is also presented in this article. The model provides a content-aware representation-and-organization platform for multimedia data manipulation in mobile heterogeneous environments. The feasibility of this model is also investigated, and compared and contrasted against several models in the literature.

Table of Abbreviations

Introduction

1.1   Mobile Multimedia: Promises and Challenges

Advances and new standards in computer visualization, image processing, and wireless communication have stimulated the explosive development of multimedia applications. These applications usually consist of large volume of multimedia data whose semantic contents cannot be represented efficiently using the traditional database models. In contrast with traditional text-based systems, multimedia applications usually incorporate much more powerful descriptions of human thought – video, audio, and images [1]. Moreover, the large collections of data in multimedia systems make it possible to resolve more complex data operations such as imprecise query or content-based retrieval. For instance, the image database systems may accept a sample picture and return the most similar images to that sample [2]. However, the conveniences of multimedia applications come with challenges to the existing data management schemes:

- Multimedia applications generally require more resources; however, the storage space and processing power are limited in many practical systems, e.g., mobile devices and wireless networks [3]. Due to the large data volume and complicated operations of multimedia applications, new methods are needed to facilitate efficient representation, retrieval, and processing of multimedia data while considering the technical constraints.

- There is a gap between user perception of multimedia entities and physical representation and access mechanism of multimedia data. Computer users often browse and desire to access multimedia data at the object level ("entities" such as human beings, animals, or buildings). However, the existing multimedia retrieval systems tend to represent multimedia data based on their lower-level features ("characteristics" such as color patterns and textures), with little regard to combining these features into objects [4]. This representation gap often leads to higher processing cost, lower throughput, and unexpected retrieval results. The representation of multimedia data according to human's perspective is one of the focuses in recent research activities; however, no existing systems provide automated identification or classification of objects from general multimedia data collections [5].

- The collections of multimedia data are often diverse and poorly indexed [6]. In a distributed environment, because of the autonomy and heterogeneity of data sources, multimedia data objects are often represented in heterogeneous formats [1]. The difference in data formats further leads to the difficulty of incorporating multimedia data objects under a unique indexing framework.

- Last but not least, the present research on content-based multimedia retrieval is based on feature vectors — features are extracted from audio/video streams or image pixels, empirically or heuristically, and combined into vectors according to the application criteria. Because of the application-specific multimedia formats, this paradigm lacks scalability, accuracy, efficiency, and robustness [1].

Based on the aforementioned discussion then one can conclude that the research on mobile multimedia technologies is currently of increasing importance in computer society. While the earlier multimedia database systems deal with single-modal data (e.g. images), most recent research focuses on providing cross-modal content-based mobile multimedia retrieval. Hence, the existing multimedia information retrieval systems (MIRS) can be divided in two categories: single-modal MIRS and cross-modal MIRS.

The single-modal MIRS focus on providing content-based indexing within a single modality, such as image or audio. Most previous research topics in this area were highly domain-specific applications (such as face recognition [30]), with less emphasis on general-purpose multimedia data manipulation [31]. Moreover, the feature-based schemes employed in earlier research (e.g. tree-based indexing models) needed large number of features and sometimes user feedback to achieve the required accuracy.

The cross-modal MIRS obtain multimedia data contents by fusing visual-audio data and camera operations into an integrated data system [32]. The recent research has focused on context-extraction models (Latent Semantic Index, Canonical Correlation Analysis, and etc.) to obtain semantic contents from video frames [33]. However, the extraction of semantic contents is a time-consuming task that includes complex matrix computations [33].

1.2   Fundamental Issues in Mobile Multimedia

Successful storage and access of multimedia data, especially in a mobile and heterogeneous database environment, requires careful analysis of the following issues:

- Efficient representation of multimedia data objects in databases,
- Proper indexing architecture of the multimedia databases, and
- Proper and efficient technique to browse and/or query data objects in multimedia database systems.

Multimedia representation focuses on efficient and accurate description of content information that facilitates multimedia information retrieval. Various approaches have been proposed to map multimedia data objects into computer-friendly formats [1, 5, 18]; however, most of the proposed schemes cannot guarantee accuracy when performing content-based retrieval. Consequently, the key issue in multimedia representation is the tradeoff between accuracy and efficiency.

As noted in the literature multimedia information retrieval methods can be classified into three groups: query-by-keyword, query-by-example, and query-by-browsing [38]. Each method is suitable for a specific application domain. Naturally, different query methods also require employment of different indexing models. Query-by-browsing is suitable for the novice user who has little knowledge about the multimedia data objects, so the aim of indexing is to compress multimedia data objects into small icons (compact representation) for fast browsing. For query-by-keyword, the keywords involve the complete semantic information about the multimedia data objects; thus it is suitable for application domains where the users have clear knowledge about the representation of the data. In the case of query-by-example, users are more interested in the effectiveness of the query processing in locating the most related images (nearest neighbors) to the given examples.

Among the aforementioned fundamental issues, multimedia representation provides the foundation for indexing, classification, and query processing. The suitable representation of multimedia entities has significant impact on the efficiency of multimedia indexing and retrieval [6]. For instance, object-level representation of multimedia data usually provides more convenient methods in content-based indexing than pixel-level representation [5]. Similarly, queries are resolved within the representation domains of multimedia data, either at object-level or at pixel-level [4]. The nearest-neighbor searching schemes are usually based on the careful analysis of multimedia representation — the knowledge of data contents and organization in multimedia systems [7].

1.3   Outline of This Chapter

The remaining part of this article is divided into three major sections. In section 2, we introduce the concepts of mobile multimedia content representation and analyze the representation approaches proposed in the literature. After examining the rationales of these approaches, their strengths and weaknesses are compared and contrasted. Finally, the semantic representation of multimedia data objects is discussed.

In section 3, we study two classes of content-based multimedia indexing models: partition-based models and region-based models. The discussion starts from the properties and rationales of the indexing models and extends to the performance analysis based on parameters such as storage utilization, searching efficiency, and robustness. Since these indexing models are all based on feature vectors, the restrictions of feature representation are also discussed.

Section 4 introduces the semantic analysis methods for multimedia data objects. By studying the proposed models on content recognition and semantic analysis, the concepts of obtaining semantics from multimedia data is addressed and analyzed. Based on these observations, a semantic-based multimedia manipulation platform is introduced. This model is a content-aware platform that provides precise and concise representation of multimedia data objects. We also investigate the feasibility of this model. In addition, the proposed model is compared and contrasted against several models as advanced in the literature. Finally, the article is concluded in section 5.

Multimedia Representation Techniques

The main goal of mobile multimedia representation is to obtain a concise description of the contents during the analysis of multimedia data objects. Representation approaches presented in the literature can be classified into four groups: clustering-based approach, representative-region-based approach, decision-tree-based approach, and annotation-based approach.

2.1   Clustering-Based Approach

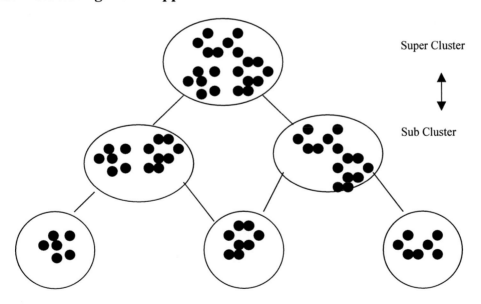

Figure 1: The cluster hierarchy.

In the clustering process, multimedia data objects (say, images) based on their semantic contents are grouped into clusters — a cluster is a group of objects with similar contents. Two types of clustering approaches have been proposed: supervised and unsupervised clustering schemes [5]. The supervised clustering approach utilizes the user's knowledge and input to cluster image data objects, so it is not a general purpose approach. As expected, the unsupervised approach does not need the interaction with user. Hence, it is an ideal method to cluster unknown image data automatically. Because of the advantage of unsupervised approach, this article focuses on unsupervised clustering schemes. It should be noted that a collection may be decomposed into several sub-clusters and represented as the union of sub-clusters (Figure 1). In this hierarchical organization, each node represents a cluster of data objects with common attributes. In addition, each node (cluster) is the union of its children nodes (sub-clusters). An indexing structure based on the attributes of the generated clusters would allow one to develop an efficient and effective retrieval scheme.

Preparation

The preparation step extracts basic features from image data objects: an image data object is partitioned into several visual component groups based on a set of given heuristics. The aim of partitioning is to ensure that the semantic distance between any two components in different groups is as large as possible, while the semantic distance between any two components in the same group is as small as possible. The partitioning process can be divided into two phases: pixel pattern recognition and feature extraction.

In the pixel pattern recognition phase, the relationship between pixels is recognized. It is assumed that the pixels in close vicinity and similar intensity values are related to the same visual component [7]. The related pixels form some pixel patterns. If this assumption holds, recognition is a process of converting pixel patterns to visual components.

Clustering

The clustering process groups the semantically similar components into clusters. Each cluster has a prototype, which is the center of the cluster, expressing the common characteristics of the visual components in that cluster. The prototypes function as the identification of their clusters.

The clustering process works in the following manner: Initially, the clustering system employs a set of randomly selected visual components as initial cluster prototypes. As expected, the randomly selected initial prototypes may not be the optimal, as a result, during the course of operations, the clustering algorithm, iteratively, attempts to refine the cluster prototypes. The ultimate goal is to obtain a generic description of clusters after several refinement processes.

2.2   Representative-Region-Based Approach

The representative-region-based approach selects several of the most representative regions from a multimedia data object and constructs a simple description of the object based on the selected regions [10] — The representative regions are small areas with the notable characteristics. In case of an image, the representative regions can be areas that the color changes markedly, or areas that the texture varies greatly, etc.

The representative-region based approach is performed as a sequence of three steps:

- Region selection,
- Function application, and
- Content representation.

2.2.1 Region Selection

The original image data object consists of many small regions. Hence the selection of representative regions is the process of analyzing the changes in those small regions. For each small region, its color, intensity, texture, and other features are compared with its neighboring regions. The difference with the neighboring regions is quantified as a numerical value to represent a region. Finally, based on such a quantitative value, regions are ordered, and the most notable regions are selected according to the following two rules:

1. The values of the regions should be as large as possible,
2. The difference between selected regions should be as large as possible.

2.2.2 Function Application

The foundation of the function application process is the Expectation Maximization (EM) algorithm [12]. The EM algorithm is used to find the maximum likelihood function estimates when the image data object is represented by a small number of selected regions. The EM algorithm is divided into two steps: E-step and M-step. In the E-step, the features for the unselected regions are estimated. In the M-step, the system computes the maximum likelihood function estimates using the features obtained in the E-step. The two steps alternate until the functions closely approximate the original features in the unselected regions.

In short, the representative-region based approach employs several representative regions and functions to construct a simple description of the original image data object. It should be noted that one may ignore some trivial details to facilitate faster processing. However, as reported in the literature, the overall accuracy of this approach is acceptable [11].

2.3 Decision-Tree-Based Approach

The decision tree model is mostly applicable in application domains where decision rules can be used as standard facts to classify the image data objects. For example, the satellite cloud images are categorized as rainy and cloudy according to the densities of clouds. In medical fields, the 2-D slices from magnetic resonance imaging (MRI) or computerized tomography (CT) are diagnosed as normal or abnormal according to the colors and shapes of body tissues.

This approach constructs the decision tree in a two-phase process: decision rule generation and decision tree construction.

2.3.1 Decision Rule Generation

The statistical data of the image features of a set of predefined image data objects (i.e., training samples) are analyzed to obtain the decision rules. Consequently, the selection of the training samples is very important to the success of this approach.

Selection of the training samples: The training samples are image data objects whose semantic information is already known. By analyzing a large number of training samples, one can find the relationship between the image features and their semantic contents. The relationships are stored as decision rules. For example, the weathercast only focuses on temperature, the density of clouds, the texture of clouds, and the ratio of dark cloud to white cloud. An example of some training samples of a weathercast is shown in Table 1.

Finding the decision rules: The decision rules are automatically deduced by analyzing the training samples. Generally, the decision rules are in the following form:

$$\{\text{feature in equation}\} \rightarrow \{\text{conclusion}\}$$

For example, from the entries in table 1, one can deduce the following decision rule:

$$(\text{temperature} \leq 45 \text{ and cloud density} > 80) \text{ or } (\text{texture} = \text{whirlpool}) \rightarrow \text{rainy condition}$$

Table 1: The extracted features of cloud images.

TEMPERATURE	CLOUD DENSITY	TEXTURE	WEATHER
45	90	plain	rainy
50	60	whirlpool	rainy
65	75	plain	sunny
38	80	plain	rainy
77	50	plain	sunny
53	85	plain	rainy
67	100	whirlpool	rainy

2.3.2   Decision Tree Construction

The decision tree construction is based on the generated decision rules. The decision rules are intended to filter out the unrelated image data objects from the pool of the available images. The fast construction of the decision tree relies on the selection of optimal decision rules that are directly related to the images. Figure 2 depicts the decision tree of rainy image data objects based on the aforementioned Rule.

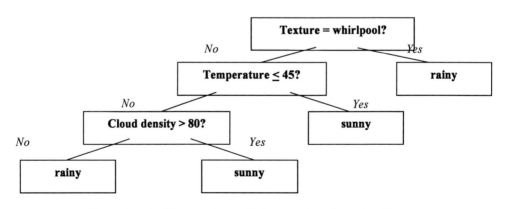

Figure 2: A decision tree for predicting weathers from cloud images.

2.4   Annotation-Based Approach

Annotation is the descriptive text attached to multimedia data objects. Traditional multimedia database systems employ manual annotations to facilitate content-based retrieval [25]. Due to the explosive expansion of multimedia applications, it is both time-consuming and impractical to obtain accurate manual annotations for every multimedia data object [1]. Hence, automated multimedia annotation is becoming an active research interest [17].

The annotation-based approach generates content in two steps: First, the concrete visual components in the multimedia data objects are identified as the basis for annotations. For instance, each image in figure 3 has three keywords indicating its visual components. Pattern recognition or object detection techniques can be used to identify basic visual components. Second, based on the basic annotations, the system deduces certain high-level semantic concepts, and keeps the abstract concepts as deducted annotations.

building, sky, cloud   car, driver, van   space ship, astronaut, flag

Figure 3: Annotation examples with three keywords.

2.4.1   Basic Annotations

The basic annotation is generated in the following way: the system recognizes the pixel patterns in the image data object and converts them into visual components. Then the visual components are replaced with the descriptive text (basic annotation of the image data object). Subsequential processing is based on the saved annotations, ignoring the visual components — this offers a high degree of efficiency since processing of text is much easier than that of pixel patterns.

Because the realization of the annotation-based approach depends on the accuracy of annotations, the basic annotations must be exhaustive, accurate and concise. Recent technological advances in image segmentation and object recognition can achieve impressive results from a large collection of images, which provides the foundation for automatic annotation extraction [21].

2.4.2   Deducted Annotations

Semantic analysis can be employed in annotation-based approach to obtain extended content description from multimedia annotations. For instance, an image containing "flowers" and "smiling faces" may be properly annotated as "happiness".

In conclusion, the annotation-based approach: (i) automatically recognizes the pixel patterns of the image data object and converts them into basic annotations in the text form, (ii) generates the deducted annotations, which reveal the implication of the basic annotations. The deducted annotations greatly enlarge the searching scope and enhance the capability of the search engine. However, as noted in the literature [11], this approach is still far from being perfect. A semantic concept may be deduced in many different combinations.

2.5   The Comparative analysis of Representation Approaches

The rationales of the aforementioned multimedia-representation approaches lead to their strengths and weaknesses in different application domains. In this sub-section, these approaches are compared under the consideration of various performance merits, such as efficiency, accuracy, storage requirement, reliability, and application domains.

2.5.1   Analysis of Clustering-Based Approach

The clustering-based approach is a great advance towards the automatic recognition of image content. As one expects, the manual recognition of image content is a tedious and time-consuming process, and the result is often hampered by the observation angels, texture, color patterns, or shapes. In general, the clustering-based approach:

- Is comprehensive, exhaustive, reliable, and accurate. It analyzes the multimedia data objects at the pixel level. As a result, this approach has the capability of revealing the details in the multimedia data objects [5].
- Is a general-purpose and easily to implement. Since the representation through clusters is straightforward, the clustering model does not rely on the knowledge of the specific domains [17].
- Is not cost effective due to its exhaustive nature. In addition, it does not have the ability to distinguish the "important" features from the "trivial" features in an object [18].
- Is not space-efficient. This is because of the size of the generated cluster hierarchy and redundancy of its contents [19].

2.5.2   Analysis of Representative-Region-Based Approach

The representative-region-based approach:

- Is faster and more space-efficient than the clustering. The high efficiency of this approach is also of interest in dynamic environment where the data objects change frequently — this model responds to the change immediately by reselecting representative regions [10].
- Is easy to implement and hence, it is widely used in different application domains [11].
- Lacks accuracy and hence, not suitable for the application domains that require the high accuracy of multimedia representation [11].

2.5.3   Analysis of Decision-Tree-Based Approach

The decision tree approach:

- Is robust. Consequently, the noise in a multimedia data object will not affect the accuracy and reliability [1].
- Has self-learning capability. The relevance feedback process allows the approach to improve the decision rules and hence to improve the accuracy of the operation [4].
- Is highly efficient. Its high efficiency lies in two aspects. First, it only analyzes a few features that affect the result. Second, it can automatically select the optimal decision rules that ensure the minimum height of the decision tree [30].
- Is not for general application domains and has found its application for some special fields [4].
- Lacks reliability. Accuracy can be improved by increasing the number of training samples. However, this would affect its efficiency [31].

2.5.4   Analysis of Annotation-Based Approach

The annotation-based approach:

- Is fast, since the traditional feature-based analysis is replaced with the text processing and there is a vast amount of established and advanced research on efficient processing of text [7].
- Is space efficient, since annotations are the concise descriptions of the multimedia contents — the multimedia data object is represented by a few text. Compared with other approaches, the space requirements of this approach is quite small [1].
- Lacks reliability, since its accuracy depends on the accuracy of annotations. If the basic annotations cannot represent the overall contents of the visual components, then the approach cannot describe multimedia data objects accurately [7].
- Is not easy to implement. It needs the support of a large annotation database. Consequently, it is not widely used.

Table 2 summarizes the aforementioned analysis and compare and contrast different classes against each other.

Table 2: Comparisons of representation approaches.

PERFORMANCE MERIT	CLUSTERING	REPRESENTATIVE REGION	DECISION TREE	ANNOTATION
Rationale	Searching pixel-by-pixel, recognizing all details	Selecting representative regions	Treating annotations as multimedia data contents	Using annotations as standard facts
Reliability & Accuracy	Reliable and accurate	Lack of robustness	Depending on the accuracy of annotations	Robust and self-learning
Time Complexity	Exhaustive, very time consuming	Most time is spent on region selection	Fast text processing	Time is spent on decision rules and feedback
Space Complexity	Large space requirement	Relatively small space requirement	Very small storage needed	Only need storage for decision rules
Application domain	Suitable for all application domains	The objects that can be represented by regions	Need annotations as basis	Restricted to certain applications
Implementation complexity	Easy to classify objects into clusters	Difficult to choose proper regions	Easily obtaining content from annotations	Difficult to obtain proper decision rules

The traditional approaches discussed above do not consider the semantic contents of the multimedia data objects. Hence, they can be referred to as "non-semantic-based" approaches. In general, these content-representation approaches:

- are not easily understood [1].
- are not robust and scalable. Each approach is suitable for some specific application domains, and hence do not offer a general solution for heterogeneous data sources.

The "semantic-based" approaches are discussed in section 4.

Content-Based Indexing

3.1   Preliminaries

The idea of content-based retrieval of multimedia data derives from the nearest-neighbor queries in multi-dimensional space. The correlation between similarity search and content-based retrieval is based on the feature-vector representation of multimedia data objects. The operations involved in content-based retrieval are defined as follows:

Definition 1: Semantic Distance

Suppose $I = \{I_j \mid 1 \leq j \leq n\}$ is the set of multimedia data objects, and $\boldsymbol{\Phi} = \{\varphi_i \mid 1 \leq i \leq m\}$ is an ordered vector of feature priorities. The semantic distance basing on feature φ_i is a function $g^{\varphi_i}: I \times I \rightarrow R$, where R is the set of real numbers. The semantic distance function g^{φ_i} compares two multimedia data objects and returns their semantic distance.

For multimedia data objects x, y, and z $\in I$, the function g^{φ_i} satisfies the following characteristics:

- $g^{\varphi_i}(x, y) \geq 0,$ (1)

- $g^{\varphi_i}(x, y) = 0 \; iff \; x = y,$ (2)

- $g^{\varphi_i}(x, y) = g^{\varphi_i}(y, x).$ (3)

- $g^{\varphi_i}(x, y) + g^{\varphi_i}(y, z) \leq g^{\varphi_i}(x, z).$ (4)

The *semantic distance* provides a quantized measure of comparing the difference between multimedia data objects. Based on the definition of semantic distance, we introduce the nearest neighbor concept that is widely used in most multimedia retrieval systems.

Definition 2: The 1-Nearest Neighbor

Assume $I = \{I_j \mid 1 \leq j \leq n\}$ is the set of multimedia data objects, $\boldsymbol{\Phi} = \{\varphi_i \mid 1 \leq i \leq m\}$ is the ordered mask of feature extraction priorities, $W = \{w_i \mid 1 \leq i \leq m\}$ is the set of weights of the feature extraction priorities, and X is the multimedia data object that is used as the query example. The nearest-neighbor searching process is a function Q:

$$Q(X, I, \boldsymbol{\Phi}, W) = \{I_i \mid I_i = min\{\textstyle\sum_{k=1}^{m}(g^{\varphi_k}(X, I_j) * w_k)\}_{j=1}^{n}\} \tag{5}$$

The searching sphere of querying
example image

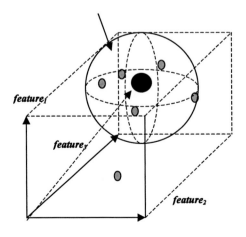

Figure 4: Search sphere for nearest neighbors.

Definition 3: The *k*-Nearest Neighbors

Assume $I = \{I_j \mid 1 \leq j \leq n\}$ is the set of multimedia data objects, $\Phi = \{\varphi_i \mid 1 \leq i \leq m\}$ is the ordered mask of feature extraction priorities, $W = \{w_i \mid 1 \leq i \leq m\}$ is the set of weight of the feature extraction priorities, k is the parameter indicating the number of nearest neighbors, and X is the multimedia data object that is used as the query example. The K-nearest-neighbor searching process is a function Q^*:

$$Q^* (X, k, I, \Phi, W) = \{I_i \mid |Q^* (X, k, I, \Phi, W)| = k, \; \forall I_j' \notin Q^*(X, k, I, \Phi, W),$$
$$\sum_{k=1}^{m}(g^{\varphi k} (X, I_i)* w_k) \leq \sum_{k=1}^{m}(g^{\varphi k} (X, I_j')* w_k)\} \tag{6}$$

Based on the aforementioned definitions, the semantic distance between multimedia data objects is quantified as the spatial distance between vertices in the feature space. In another word, the nearest neighbors resides within a sphere whose center is the querying image data object (Figure 4).

3.2   Content-Based Indexing Models

Index is an infrastructure that facilitates access to the data. It works as the interface between the user and the multimedia database. The index fetches the proper multimedia data objects from the database and returns them as query results. The content-based multimedia retrieval models that have advanced in the literature can be categorized into two classes: partition-based models and region-based models.

3.2.1   Partition-Based Models

The partition-based indexing models recursively divide the multimedia data objects (or *k*-dimensional feature spaces) into disjoint partitions, with clustering or classifying algorithms, while generating a hierarchical indexing structure on these partitions. The earlier models in this class include Quad-tree [12], K-D-tree [13], and VP-tree [14].

Quad – Tree

A quad-tree has a tree-shaped indexing structure that is based on the recursive decomposition of multidimensional space. On the root of the quad-tree, the multidimensional space is decomposed into four regions, denoted as NW, NE, SW and SE. And these four regions are decomposed recursively.

Originally quad-tree was defined to represent irregular regions, i.e., region quad-tree. Another type of quad-tree is point quad-tree. It is used to represent the points in multidimensional space, such as cities in a map. A quad-tree normally has the following characteristics:

- The quad-tree has a hierarchical data structure: In a quad-tree, each node (except the leaf nodes) has four children nodes. Each child node represents a sub region decomposed from its parent [12].

- The quad-tree representation is not unique: In the decomposition process, the four sub regions are selected according to the strategies chosen in real applications. Figure 6 depicts a quad-tree generated by selecting data points in Figure 5 according to alphabetic order. Based on different strategies, the same set of data objects may be decomposed into different sub regions.
- The quad-tree is not a balanced tree: The quad-tree is constructed during the decomposition process. The image data objects are recursively decomposed until the sub regions satisfy the requirement.
- The search time and space requirement of a quad-tree depend on the number of data objects (or leaf nodes) in the tree. For a quad-tree with n leaf nodes, the average search time for nearest-neighbor queries is $O\ (n \log n)$, and the space required for performing the retrieval is $O\ (n^2)$.

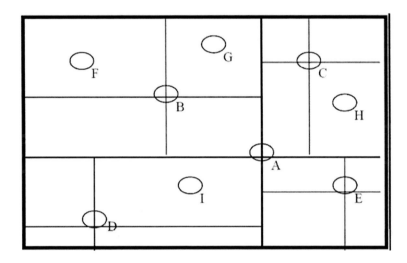

Figure 5: Data points in a 2-dimensional space.

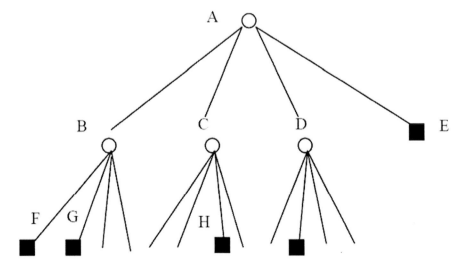

Figure 6: An example of quad-tree.

K-d – Tree

K-d tree is a multidimensional binary search tree [13]. The parameter k denotes the dimensionality of the search space. It was originally designed to facilitate the multi-key retrieval in a multidimensional space. Similar to quad-tree, the k-d tree also has the capability of indexing multimedia data objects in multi-dimensional space. Figure 8 depicts the k-d tree obtained from a set of 2-d points in Figure 7. A k-d tree normally has the following characteristics:

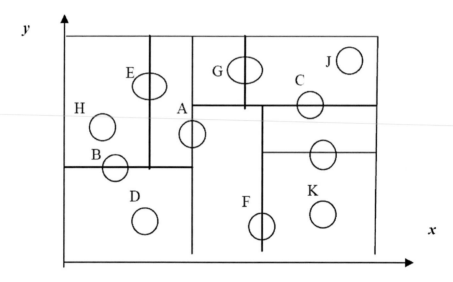

Figure 7: The points in a 2-D space.

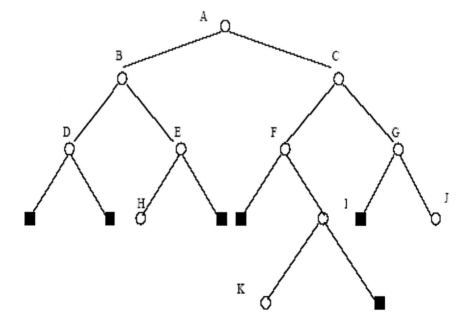

Figure 8: An example k-d tree.

- A *k-d* tree is an unbalanced binary search tree: The structure of *k-d* tree depends on the points and their selection order. There is no balancing operation in the selection process.
- The *k-d* tree can accept multiple query forms: The *k-d* tree has the capability of handling many queries efficiently, including exact match query, partial match query, insertion query, region query and nearest neighbor query [13].
- The time efficiency of *k-d* tree is as follows [13]: For a system with *n* data objects, the average insertion time is $O(log\ n)$, the average deletion time of a random node is $O(log\ n)$. The deletion time of the root node is $O(n^{(k-1)/k})$. If *t* keys are given for a partial match query, the average searching time is $O(n^{(k-t)/k})$.

VP – Tree

VP-tree (vantage point tree) has an indexing structure for the nearest-neighbor retrieval in high-dimensional spaces [14]. In these spaces, the dimensionalities are so high that the traditional methods of decomposition are not efficient. So the VP-tree does not decompose the space according to their dimensions. It employs some vantage points as the standard for partitioning the space. The vantage points are some distinguished points that are capable of dividing the other points evenly into some partitions.

The VP-tree construction begins with the selection of a vantage point. The vantage point is selected by computing the maximum standard deviation of distance values. The selection criteria of vantage points may vary in different applications, but their goals are the same: the vantage points should partition the other points evenly.

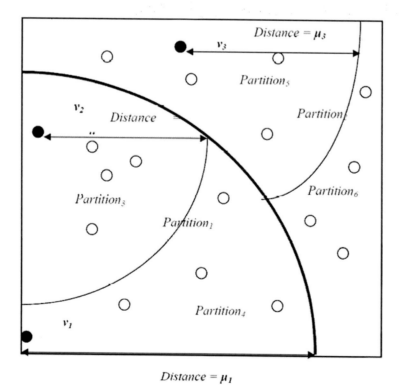

Figure 9: The partitions of space.

Figure 9 depicts the partitions of a space. The VP-tree approach first selects v_1 as the vantage point, and computes the mean of distance values (distance1). The points with distance smaller than distance1 are grouped in partition1, and the points with distance greater than distance1 are grouped in partition2. In partition1, a new vantage point v_2 is selected. The point v_2 divides partition1 into partition3 and partition4. Similarly, the vantage point v_3 divides partition2 into partition5 and partition6. Figure 10 depicts the corresponding VP-tree.

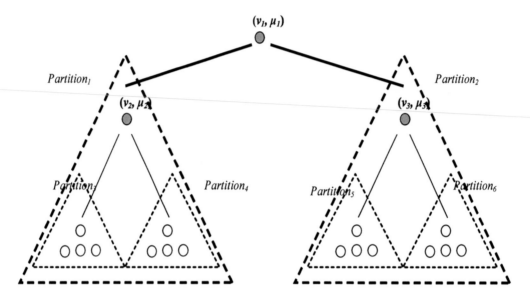

Figure 10: The VP-tree.

Based on the aforementioned rationality of VP-tree, the following properties can be summarized:

- The VP-tree has a distance-based indexing structure: In some multimedia databases, the objects may not be easily converted into feature vectors. So they cannot be considered as vertices in a multidimensional space. However, their distances might be easily obtained.
- The VP-tree facilitates the finding of nearest neighbors: The VP-tree approach employs a binary search tree to represent all the data objects. Each node in the VP-tree denotes a data object. According to the searching algorithms proposed in [14], the search scope of nearest neighbors is restricted to a few branches of the VP-tree. As a result, the search of nearest neighbors is easier.
- The time efficiency of VP-tree is quite high: The VP-tree construction time is $O(n \log n)$, and the expected searching time is $O(\log n)$. It was shown that the searching in the VP-tree is more efficient than that of the R^*-tree and the M-tree.

3.2.2   Region-Based Models

The region-based indexing models employ small regions (either in the form of Minimum Bounding Rectangles or Spheres) to cover all the multimedia data objects. This class of indexing models includes the R-tree and its variations (R*-tree, R$^+$-tree) [20], and SR-tree [21]. Relative to partition-based models, this indexing model improves storage utilization by avoiding forced splits. However, due to the shapes of the regions, this model has the weakness of overlapping [20].

R – Tree

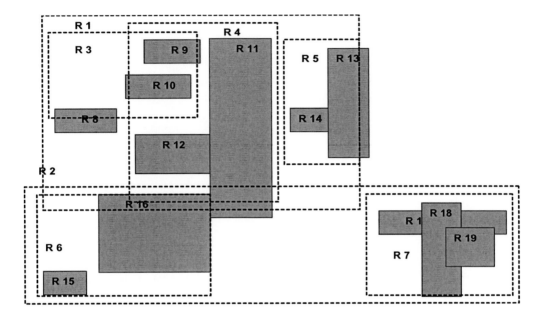

Figure 11: The data objects enclosed in rectangles [20].

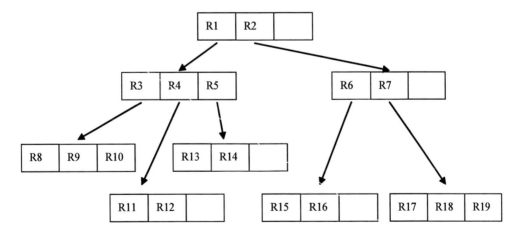

Figure 12: The R-tree [20].

The R-tree has a hierarchical indexing structure for spatial data objects. It employs rectangular regions to group the data objects. It is one of the most widely applicable indexing methods in multimedia databases. To provide the indexing on data objects, R-tree usually has the following properties:

- The R-tree uses minimum bounding rectangles (MBR) to construct the index [20]: The MBR is a kind of hyper-rectangular region in the multidimensional data space. Overlapping between different MBRs in different branches is allowed, although this overlapping lowers the efficiency of R-tree [20]. This deficiency, however, is overcome by its variations R^+-tree and R^*-tree.
- The R-tree is a balanced B-tree like structure: The R-tree uses the similar structure as B-tree. It includes leaf nodes and non-leaf nodes. All the indexed data objects are kept in the leaf nodes. The non-leaf nodes indicate the relationship among the data objects.
- The R-tree is a dynamic indexing structure: In R-tree, the insertion and deletion operations can be intermixed with searching operations, as a result, there is no need for the periodical reorganization of the R-tree [20].

The construction of R-tree is a top-down process. Starting from a root node, the R-tree grows as more data objects are added to the database. Let M be the maximum number of entries in one node. And let $m \leq M/2$ be the minimum number of entries in one node. The R-tree approach first selects a MBR that includes M data objects, and makes this MBR as the root node. And then, insert data objects one by one into this R-tree. If a node overflows, the R-tree approach splits this node into two child nodes, which represent two smaller MBRs. This insertion process continues until all the data objects are included in the R-tree. This R-tree is proven to have a height of at most $|log_m n| - 1$, where n is the number of data objects. And all the leaf nodes appear on the same level of the R-tree [22].

Figure 11 depicts the minimum bounding rectangles that enclose the data objects in a multidimensional space. R_8, ..., R_{19} denote the data objects. R_1, ..., R_7 are the MBRs that enclose the data objects. Figure 12 depicts the corresponding R-tree, in which M = 3 [22].

R^+ – Tree

The R^+-tree model is a variation of the R-tree model. It avoids the overlapping rectangles in the non-leaf nodes of the R-tree. This improvement allows the R^+-tree to save up to 50% disk access time relative to that of R-tree [20].

The R^+-tree improves the R-tree by avoiding the overlaps. However, in multimedia databases, the data objects may be concatenated together, which makes it difficult to find overlap-free rectangles. In such cases, the R^+-tree approach will force some data objects to split into smaller objects in order to find the minimum bounding rectangles at proper positions.

Figure 13 depicts the rectangles for an R^+-tree. R_8, ..., R_{19} denote the data objects. To enclose the data objects in overlap-free rectangles, the R^+-tree approach forces R_{11} to split into two parts, so it gets the MBRs R_1, ..., R_3. And then, it forces R_{12} to split, and gets the MBRs R_4 and R_5. Figure 14 depicts the corresponding R^+-tree of Figure 13.

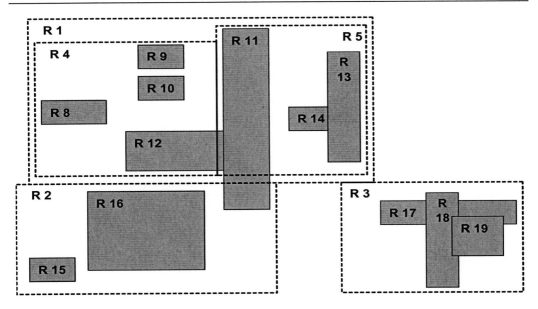

Figure 13: The rectangle regions of R⁺-tree.

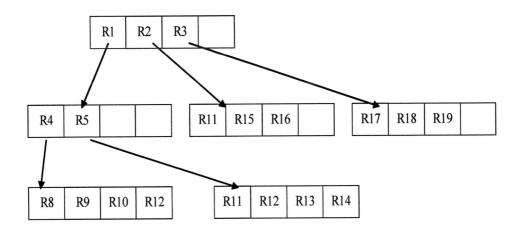

Figure 14: The R⁺-tree.

R* – Tree

The R^*-tree model is the most efficient approach in R-tree family. It does not require the MBRs to be overlap-free. It uses heuristic splitting algorithms to find the optimal MBRs, and achieves the higher performance [23]:

- R^*-tree improves the efficiency of nearest-neighbor retrieval: The R^*-tree minimizes the overlaps and avoids recursive splitting, so it has superior performance than other models in the R-tree family. According to the experimental results, the performance of R^*-tree is about four times as fast as that of R-tree [23].

- The storage utilization of the R*-tree is high: With the reduced area, margin and overlap of the MBRs, the R*-tree gets optimal utilization of the storage. This property also decreases the disk access time.
- The R*-tree is a robust variation of R-tree: The R*-tree is capable of handling all kinds of data distribution efficiently. The abnormal input data will not greatly deteriorate the performance.

SS – Tree

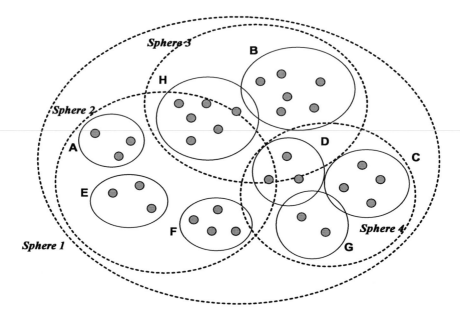

Figure 15: The spheres in a 2-D space [17].

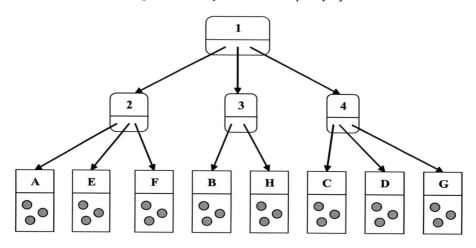

Figure 16: The SS-tree [17].

The SS-tree (similarity search tree) model has a dynamic indexing structure for the nearest neighbor retrieval (similarity comparison) in high dimensional space. SS-tree usually supports the following characteristics for the purpose of efficient indexing:

- The SS-tree employs spheres to enclose data objects: Unlike the R-trees, the SS-tree groups the data objects using spheres. For maintenance convenience, the spheres do not need to be minimum bounding spheres [21]. In SS-tree, a sphere is represented by its centroid (the mean value of the data objects in it). The spheres can be divided into smaller spheres when it overflows.

- The SS-tree has better performance than R-tree family: It has been theoretically proven that the sphere has smaller Minkowski sum than the volume-equivalent MBR [24]. As a result, the sphere based SS-tree achieves superior performance than the MBR based R-trees. As shown in [24], the performance of SS-tree is two times higher than that of R^*-tree.

- The SS-tree does not avoid the overlaps among the spheres: One problem of sphere is that a big sphere can not be split into two smaller overlap-free spheres. So the SS-tree allows overlapping spheres. The SR-tree model overcomes this problem.

Figure 15 depicts the spheres in a 2-dimensional space. A, ..., H are the spheres that enclose the data objects. These spheres are leaf nodes in SS-tree. The sphere 1,.., 4 enclose the smaller spheres, which refer to the non-leaf nodes in SS-tree. Figure 16 depicts the corresponding SS-tree.

SR – Tree

The SR-tree (sphere/rectangle-tree) model is the combination of SS-tree model and R^*-tree model. It employs both spheres and rectangles for indexing data objects (Figure 17). The performance improvement of SR-tree is achieved through the advantages gain from both spheres and rectangles partitioning:

- The SR-tree has better storage utilization [21]: To enclose data objects, the spheres occupy more than necessary space and cause the overlaps among spheres. The SR-tree uses rectangles to get the overlap-free splitting. As a result, the SR-tree greatly reduces redundancy in the indexing structure and hence increases the storage utilization.

- The SR-tree improves indexing performance: The spheres are efficient in handling nearest neighbor indexing and range indexing. In addition, the rectangles are more suitable in dealing with overlap-free splitting. The SR-tree inherits the advantages of SS-tree and R*-tree, so it outperforms both of them.

SS-Tree   R*-Tree   SR-Tree

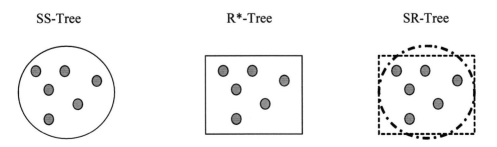

Figure 17: The comparisons of SS-Tree, R-Tree, and SR-Tree.

3.3   Comparisons of Content-Based Indexing Models

Table 3: Properties of content-based indexing models.

NAME	INDEXING FOUNDATION	BALANCED	BINARY TREE	OVERLAP	RE-INSERT	HEURISTIC ALGORITHM
Quad tree	NW, NE, SW, SE	No	No	No	No	No
k-d tree	k dimensions	No	Yes	No	No	Yes
VP Tree	vantage point	No	Yes	No	No	Yes
R-tree	MBR	Yes	No	Much	No	Yes
R*-tree	MBR	Yes	No	Little	Yes	Yes
R⁺-tree	MBR	Yes	No	No	No	Yes
SS tree	sphere	Yes	No	Much	Yes	Yes
SR tree	sphere, MBR	Yes	No	Little	Yes	Yes

Table 4: Performance comparisons of content-based indexing models.

NAME	HIGH DIMENSIONAL	QUERY TYPE	STORAGE UTILIZATION	ACCESS EFFICIENCY	DATA OBJECT	DYNAMIC ORGANIZATION
Quad tree	Poor	Region	Poor	Low	Point	No
k-d tree	Fair	Region, NN	Poor	Low	Point	No
VP tree	Good	Range, NN	Poor	High	Point	No
R-tree	Poor	Region, range,NN	Poor	Low	Area	Yes
R*-tree	Good	Region, range,NN	Good	High	Area	Yes
R⁺-tree	Poor	Region, range,NN	Poor	Low	Area, Point	Yes
SS-tree	Good	Range, NN	Good	High	Point	Yes
SR-tree	Good	Region, range,NN	Good	High	Area, Point	Yes

The aforementioned content-based indexing models have both strengths and weaknesses, when considered for different application domains. Their properties are compared and contrasted in Tables 3 and 4.

Semantic Content Manipulation

4.1 Preliminaries

As more and more multimedia data repositories are built in many applications, accurate content accessing strategies are needed for efficient retrieval of multimedia data objects. An ideal multimedia information retrieval system should have the capability of understanding, not just similarity matching, of multimedia data objects. For instance, when seeing a talking face we expect to visualize that he/she is expressing, the sound of car engine usually comes with the image of a running car.

To realize the multimedia data manipulation at the semantic level, one needs to recognize and represent the multimedia data as semantically meaningful objects. Most existing systems achieve this goal in three steps: object segmentation, object recognition, and semantic analysis.

4.2 Multimedia Data Object Segmentation

The main goal of multimedia data object segmentation is to distinguish objects from background. For instance, the pixels in an image may have different intensities, which cause the human perception of objects. If the pixels can be automatically classified into different intensity groups, the whole image is then divided into segments. The proposed segmentation algorithms can be classified into three categories: threshold-based approaches, edge-based approaches, and connectivity-based approaches [28].

4.2.1 Threshold-Based Approaches
In this approach, an intensity threshold θ is defined and applied to every pixel. If the pixel has an intensity value smaller than θ, then recognize this pixel as belonging to the object, and set the brightness as 0 (black color). Otherwise, the pixel is determined as "non-object", and is set to 1 (white color). Hence, the segmentation result is a binary-color image, in which the objects are shown as black-color regions (figure 18).

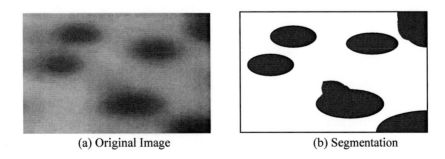

(a) Original Image   (b) Segmentation

Figure 18: An example of threshold-based approach.

The key issue in threshold-based segmentation is the determination of the threshold value. There are two ways to choose the threshold: The first method defines the threshold as the median value of intensity span. The second method determines the threshold from analysis of intensity histogram. Normally, the boundaries between objects and background exist at the point that separates smooth curve and big fluctuation (figure 19).

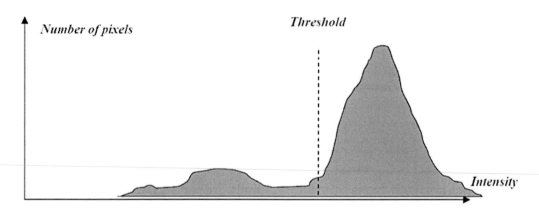

Figure 19: Selecting threshold from histogram.

4.2.2   Edge-Based Approaches

The edge-based approach is based on detecting the intensity contours in the images. The pixels within these contours are identified as "object", while the outside pixels are considered as "non-object background". In figure 20, the pixels of same intensities form the contours enclosing the objects. However, these contours are required to be enclosed to ensure accurate segmentation.

(a) Original Image   (b) Segmentation

Figure 20: An example of threshold-based approach.

The main weakness of edge-based segmentation is the failure to deal with disconnected contours. Since it does not provide an enclosed region, the segmentation algorithm cannot determine which pixels are considered as the "object". In addition, this approach does not perform well when dealing with noises [29].

4.2.3   Connectivity-Based Approaches

The connectivity-based approach works as follows: It first finds a contour as the initial boundary curve. Then iteratively modify this contour by applying various shrink/expansion operations. Finally, the shape of this contour is determined, and the pixels within this contour are identified as the object (figure 21).

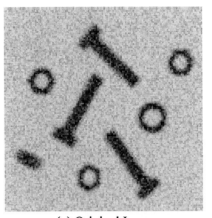   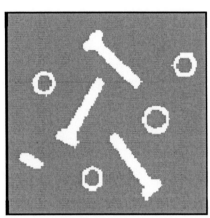

(a) Original Image   (b) Segmentation

Figure 21: An example of threshold-based approach [29].

Table 5 summarizes the aforementioned object segmentation approaches and contrasts the characteristics of these approaches:

Table 5: Comparisons of segmentation approaches.

CATEGORY	HIGH DIMENSIONS	ACCURACY	IMPLEMENT	ROBUSTNESS
Threshold-Based	Good	Fair	Good	Good
Edge-Based	Fair	Poor	Fair	Poor
Connectivity-Based	Poor	Good	Good	Fair

4.3   Object Detection and Recognition

In multimedia information retrieval systems, features can be recognized at two levels: granule-level features and object-level features. The granule-level features are those characteristics that are derived directly or indirectly from the original format of multimedia storage — i.e., the pixels, such as hue, textures, and saturation. The object-level features, in contrast, are obtained from the recognition of the higher-level understanding of the multimedia data — the semantic topics of the multimedia data [34].

Most existing multimedia retrieval systems tend to use granule-level features for content representation. This is mainly due to the difficulty of obtaining object-level semantic concepts from multimedia raw data [31]. However, human's perception of multimedia data is mostly at the object-level, instead of focusing on granule-level features. Hence, research on object detection is becoming important in recent years, and many models were proposed for efficient extraction of objects from multimedia data [35].

Based on domain-specific knowledge or empirical observations, several experimental systems were built for detection of certain types of multimedia data objects, such as face identification from images or video segments [32]. As noted in section 3.1, a multimedia system usually maps multimedia data objects to vertices in a high-dimensional space of granule-level features. A given multimedia data object may reside in a small region, surrounded by other objects that are semantically similar to it. Larger distance between vertices in the feature space corresponds to smaller content similarity between multimedia data objects (Figure 22).

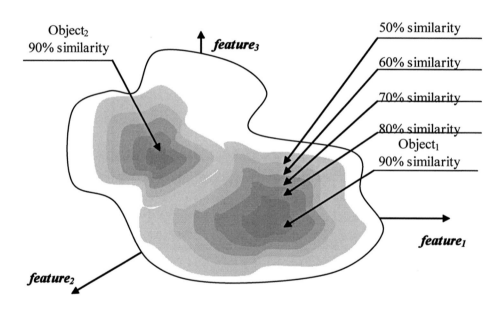

Figure 22: Object detection in feature space.

Many object recognition applications deal with objects of fixed shapes, e.g. face detection and digit recognition. The SVM is a learning machine that can perform binary classification of multimedia data [30]. It first maps the input multimedia data objects as data points in feature spaces. Then the SVM assigns these data points into one of two disjointed spaces. One is the input feature space in which a linear classifier is constructed; the other is a higher dimensional space in which a non-linear classifier is constructed.

Some applications require recognition of more flexible shaped objects, such as moving animals. One of the popular flexible-shape object recognition models is Hidden Markov Probability Model (HMM) model [25]. The HMM is a finite set of states, each of which is associated with a probability distribution. Transitions among the states are governed by a set of probabilities called transition probabilities. These probabilities are determined by the training data during the construction of the HMM.

4.4   Semantic Analysis of Multimedia Data Contents

Object recognition provides the foundation for the semantic analysis of multimedia data contents. The existing semantic-based models manipulate multimedia data by providing a cross-modal understanding at the conceptual level, instead of only involving low-level features. Depending on the application domains, the cross-modal semantic-analysis methods can be categorized as two types [33]:

    (i)   Model-dependent approaches, which employ content correlation models such as Gaussian distribution or linear correlation [36], and

    (ii)   Model-free approaches, which require little prior knowledge and are adaptable to most applications, such as neural-network-based approaches [37].

In practical applications, the model-dependent approaches usually require less training examples and can achieve better performance [33]. Consequently, the rest of this discussion focuses on the model-dependent approaches in this section.

4.4.1   Latent Semantic Indexing (LSI)

Latent semantic indexing (LSI) was originally proposed as a statistical information retrieval approach to discover underlying semantic relationship between different textual units [38]. LSI mainly focuses on two types of semantic relationships:

- Synonymy, which refers to the fact that many words may indicate the same object. For instance, the word "picture" can be referred to either as an "image" or a "photo".
- Polysemy, which refers to the fact that most words have more than one meaning in different contexts. For instance, "bus" may refer to as a means for public transportation when it appears near the word "road", while it may mean electronic circuits within the scope of computer architecture.

As mentioned in section 4.1, the semantic relationships also exist in multimedia applications. The feature-based multimedia systems usually employ large number of low-level features to increase the accuracy of content-based retrieval [32]; however, this method leads to both high computation complexity and incapability of manipulating semantic concepts. Most feature-vector-based multimedia systems decide the relationship between multimedia data objects simply by comparing their common features. As a result, unrelated data objects may be retrieved simply because similar feature values occur accidentally in them, and on the other hand, related multimedia data objects may be missed because no similar feature values occur in the query. The goal of LSI is to overcome these shortcomings by mapping the high-dimensional feature space to a lower dimensional "concept space", thus reduces the computation complexity while providing the semantic relationships between multimedia data objects.

A latent semantic index built on multimedia data usually employ a technique known as Singular Value Decomposition (SVD) to create the concept space. For instance, a multimedia system may try to set up the relationships between talking faces in video frames and their expressed words in the audio segments. A joint feature space with n video features and m audio features for u video frames may be expressed as follows [33]:

$$X = [V_1, V_2, ..., V_n, A_1, A_2, ..., A_m]^T \tag{7}$$

Where

$$V_i = (v_i(1), v_i(2), ... v_i(u)) \tag{8}$$

and
$$A_i = (a_i(1), a_i(2), ... a_i(u)) \tag{9}$$

The singular value decomposition can be expressed as follows:

$$\boldsymbol{X = K S D^T} \tag{10}$$

where K and D are orthonormal matrices composing of left and right singular vectors, and S is a $r*r$ diagonal matrix of singular values sorted in descending order in which $r = min(n+m, u)$. It can be proved that such decomposition always exists [33].

By selecting the k largest singular values in the decomposition result, the original $n+m$ dimensional feature space is mapped to a k-dimensional concept space. This will result in the decreased dimensionality and hence reducing the computation complexity. This would also allow related multimedia features be clustered together by being assigned to the same concept. However, this performance improvement comes at the expense of the following drawbacks [33]:

- The concept space generated by LSI is not readily understandable to humans. The concepts and their relationships are all represented as numbers without semantic meaning. Hence, it is difficult to make modification to the LSI concepts.
- The SVD algorithm requires a complexity of $O((n+m+t)^3 k^2)$. Typically, k can be a small value in practical applications, but the term $n+m+t$ needs to be large enough to guarantee accuracy. This makes SVD algorithm unfeasible for large and dynamic multimedia data collections.
- Determination of the optimal number of dimensions in concept space is a difficult problem. The updates of multimedia data collections may request the changes of concepts, since some added multimedia data objects may introduce new concepts, while some deleted multimedia data objects may result in the obsolescence of old concepts — it is quite time consuming to perform a new run of SVD algorithm. As one can conclude, LSI is not suitable for dynamically changing multimedia data collections.

4.4.2   Canonical Correlation Analysis (CCA)

In the LSI model, the features from different modalities (e.g. features from image and audio data) are treated as different dimensions, which may result in the loss of conceptual inter-relationship between modalities. To overcome this weakness, related features from different modalities need to be coupled together. The canonical correlation analysis (CCA) is a method of measuring the linear relationship between two multi-dimensional variables [36]. Since multimedia data objects can be mapped to different high-dimensional feature spaces (such as video and audio features), CCA is also useful for clustering features from different modalities.

Formally, CCA can be defined as the problem of finding two sets of basis vectors for two matrices X and Y, such that the correlations between the projections of the matrices onto the basis vectors are mutually maximized. For an identity matrix I, the goal is to find orthogonal transformation matrices A and B that can maximize the expression [36]:

$$\| XA - YB \|_F^{\ 2} \tag{10}$$

where $A^T A = I$, and $BB^T = I$ (A^T and B^T refer to the transformed matrices of A and B respectively). $\|M\|_F$ is the Frobenius norm of matrix M that can be expressed as:

$$\|M\|_F = (\sum_i \sum_j |m_{ij}|^2)^{1/2}, \tag{11}$$

where m_{ij} indicates the element on the ith row and j^{th} column of the matrix.

It has been shown that CCA outperforms LSI in matching video frames with their related audio sounds [33]. However, the linear relationships between multimedia data objects are still represented as numbers that cannot be understood by humans. Moreover, neither LSI nor CCA is capable of describing the hypernym / hyponym relationship between multimedia data objects or concepts. For instance, "vehicle" is a hypernym of "bus", "car", and "truck". These shortcomings justify the research to develop new schemes that provide more accurate description of semantic concepts.

4.5   Content-Aware Multimedia Representation

The aforementioned content-based indexing models have attempted to provide solutions to multimedia searching based on specific features [7]. However, in most practical cases, the semantic contents of multimedia data objects may not be indicated as low-level features. As pointed out by Wang et al. [25], human beings tend to view images as whole objects. This object-oriented view on multimedia content processing has led to the research on multimedia semantics manipulation. One of the promising models is the Extended Summary-Schemas Model (ESSM) [39].

4.5.1   Preliminaries

Basic Concepts

The original Summary-Schemas Model (SSM) is a content-aware organization prototype proposed to enable imprecise text queries on distributed heterogeneous data sources [26]. It provides a scalable content-aware indexing method based on the hierarchy of summary schemas, which comprises three major components: a thesaurus, a collection of autonomous local nodes, and a set of summary-schemas nodes (Figure 23).

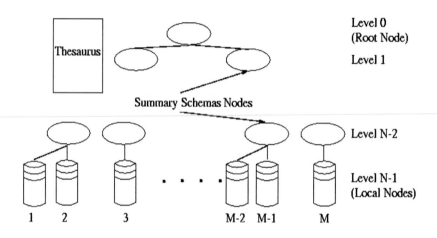

Figure 23: The Summary Schemas Model.

The thesaurus provides an automatic taxonomy that categorizes the standard accessing terms and defines their semantic relationships. The thesaurus may utilize any of the off-the-shelf thesauruses (e.g. Roget's Thesaurus) as its basis. A Semantic-Distance Metric (SDM) is defined to provide quantitative measurement of "semantic similarity" between terms [26]. A local node is a physical database containing the multimedia data. The local node may be organized by its autonomous indexing structure, on condition that the semantic contents of local data items can be easily obtained by the global accessing mechanism. With the help of the thesaurus, the data items in local databases are classified into proper categories and represented with abstract and semantically equivalent summaries. A summary schemas node is a virtual database concisely describing the semantic contents of its child (children) node(s). More detailed descriptions can be found in [26, 27].

Logic-Based Representation

To represent the contents of multimedia data objects in a computer-friendly structural fashion, the ESSM employs a mechanism to organize multimedia data objects into layers according to their semantic contents [39]. A multimedia data object, say, an image, can be considered as the combination of a series of elementary entities, such as animals, vehicles, and buildings. And each elementary entity can be described using some logic predicates, which indicate the mapping of this elementary entity on different features. For instance, the visual objects in Figure 24 are dogs and cats. The possible color is grey, white, blue, or brown. Hence, the example image in Figure 24 can be represented as the combination of visual objects, colors, and textures, such as (cat ∧ (brown ∨ blue)) ∨ (dog ∧ (grey ∨ white)).

To further provide efficient indexing on multimedia data objects, the ESSM organizes data objects based on their semantic relationships. In a distributed environment, despite the heterogeneity and autonomy of the multimedia data objects, the summary schemas represent the semantic relationships between these objects. The ESSM employs three types of links to indicate the semantic relationships:

- In a multimedia database, synonyms are either similar objects in different data formats or same-format objects in different physical locations. The ESSM employs synonym links to connect and group the similar multimedia data objects together.

- A hypernym is the generalized description of the common characteristics of a group of multimedia data objects. For instance, the hypernym of dogs, monkeys, and horses is mammal. To find the proper hypernyms of multimedia data objects, the ESSM maintains an on-line thesaurus that provides the mapping from multimedia data objects to hypernym terms. Based on the hypernyms of multimedia data objects, the ESSM can generate the higher-level hypernyms that describe the more comprehensive concepts. Recursive application of hypernym relation generates the hierarchical meta-data of the ESSM. This in turn conceptually gives a concise semantic view of all the multimedia data objects.

- A hyponym is the counter concept of a hypernym. It is the specialized description of the precise characteristics of multimedia data objects. It inherits the abstract description from its direct hypernym, and possesses its own particular features. The ESSM uses hyponyms links to indicate the hyponyms of every hypernym. These links compose the routes from the most abstract descriptions to the specific multimedia data objects.

Figure 24: Semantic content representation of image objects.

Based on the hypernym/hyponym relationships between multimedia data objects, a hierarchical indexing structure is constructed to facilitate content-based multimedia retrieval. In Figure 25, each non-leaf node captures and maintains the semantic contents of its children nodes.

One of the merits of the ESSM is the ease of nearest-neighbor search. In the ESSM, the nearest neighbors are considered as synonyms — they are connected through synonym links. As a result, searching for the nearest-neighbor is simplified into a process of finding synonyms through links. In other indexing models, the nearest neighbor indexing is a time-consuming process and needs to search a considerable section of multimedia database [1].

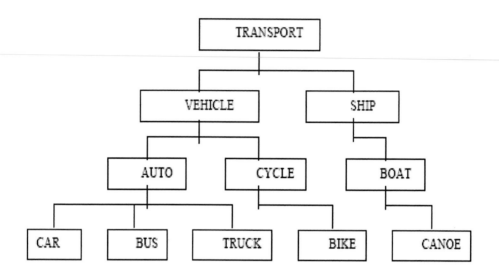

Figure 25: The ESSM hierarchy for multimedia objects.

4.5.2   Non-Semantic-Based Representation Methods vs. ESSM

In contrast with the multimedia-representation approaches mentioned in section 2, the ESSM employs a unique semantic-based scheme to facilitate multimedia representation. In the ESSM, a multimedia data object is considered as a combination of logic terms, whose value represents the semantic content. The analysis of multimedia data contents is then converted to the evaluation of logic terms and their combinations. This content representation approach has the following advantages:

- The ESSM integrates heterogeneous data sources into a unified logical system. Within the scope of multimedia databases, the data heterogeneity greatly degrades the performance — as expected, in a mobile environment, this problem becomes even more severe. Many factors contribute to the data heterogeneity. The name and data format differences are the main source of heterogeneity. For instance, the image data may be stored as BMP, GIF, JPEG, or TIFF formats. The ESSM organizes multimedia objects regardless of their physical storage formats, uniformly, according to their contents. In addition, different media types (video, audio, image, and text) can be integrated under the ESSM umbrella. In short, the ESSM organization approach avoids the complex conversion among data formats, and operates the data objects at a higher level.

- The semantic-based descriptions (logic terms) provide a convenient way to represent multimedia contents precisely and concisely. Easy and consistent representation of the elementary objects based on their semantic features simplifies the semantic content representation of complex objects using logic computations. As a result, the similarity between objects can be considered as the equivalence of their corresponding logic terms. Moreover, this logic representation of multimedia content is often more concise than feature vector, which is widely used in non-semantic-based approaches. In a specific multimedia database system, the feature vector is often fixed sized to facilitate the computation and representation [6]. However, in many instances some features may be null. Although the null features do not contribute to the semantic contents of multimedia objects, they still occupy space in the feature vectors — hence, lower storage utilization. In contrast, the logic representation can improve storage utilization by eliminating the null features from logic terms.

- The ESSM provides a mathematical foundation for operations such as similarity comparison and optimization. Based on the equivalence of logic terms, the semantically similar objects can be easily found and grouped into same clusters. This organization facilitates the nearest-neighbor retrieval, and at the same time reduces overlapping and redundancy, resulting in efficient search operation and efficient storage utilization. The optimization of semantic representation can be easily performed on logic terms using mathematical analysis. By replacing long terms with mathematically equivalent terms of shorter lengths, the multimedia representation can be automatically and systematically optimized.

4.5.3   Feature-Based Indexing Models vs. ESSM

The ESSM proposes a novel approach to content-based multimedia indexing. It possesses several particular properties in organization of multimedia data objects, which greatly improves the performance:

- The ESSM organizes multimedia objects in a hierarchical fashion that facilitates content-based retrieval (Figure 23). The lowest level of the ESSM hierarchy comprises multimedia data objects, while the higher levels consist of summary schemas that abstractly describe the semantic contents of multimedia data objects. Due to the descriptive capability of summary schemas, the height of the ESSM hierarchy is short, which drastically reduces the searching cost.

- Finding nearest neighbors in the ESSM is a direct and effective process. In the ESSM, the nearest neighbors are considered as synonyms. They are connected through synonym links. As a result, the nearest-neighbor indexing is simplified into a process of finding synonyms through links. This process can achieve the $O(1)$ time complexity in most cases.

- The ESSM provides an intelligent and friendly user interface that resolves user queries intelligently based on the contents. Moreover, with the help of on-line thesaurus, the ESSM can easily accommodate imprecise requests. In another word, the ESSM supports both precise and fuzzy queries — the user can submit various types of queries such as nearest-neighbor query, exact match query, and fuzzy query (imprecise query). For instance, the user may submit a word "nature" as query term, and the ESSM has the capability of searching for images related with specific "nature" sceneries, such as hills, trees, and rivers.

- The ESSM meta-data has the potential to be dynamic and self-adjusting. As a result, it is possible to come up with a semi-balanced hierarchical structure in which the summary-schema nodes and data objects are distributed evenly. After its creation, the summary schema's hierarchy can be dynamically modified to ensure the minimum height. For instance, in a mobile ad-hoc environment, the ESSM may be employed as a time-efficient scheme for content-based retrieval. The ESSM is capable of dynamically adjusting its structure according to the changes of network topology, which provides a reliable platform for manipulating multimedia objects in different conditions.

- Compared with many other content-based indexing models, the ESSM achieves better robustness in most cases. The feature-based models depend highly on the extraction of features to provide accurate retrieval. Some trivial errors in features may cause totally different results. In contrast, the ESSM is capable of handling all kinds of representation and distribution of data objects, and building up the efficient indexing hierarchy based on abstraction. The overall performance will not be hampered by abnormal input.

Based on the aforementioned discussions and observations, the comparisons and contrasts of semantic analysis models are summarized in Table 6.

Table 6: Comparisons of multimedia semantic analysis models.

MODEL	SEMANTIC RELATIONSHIPS	APPLICATION	MODALITY	COMPLEXITY	UNDERSTAND-ABILITY
LSI	synonomy, polysemy	content-based retrieval	unique	poor on large data collection	not understand-able to human
CCA	cross-modal	correlation analysis	multiple	fair	linear dimension relationships, not understandable
SSM	synonym, hypernym, hyponym	content-based retrieval	multiple	good in large and dynamic data sources	easily understood due to linguistic user interface

Conclusion

The content-based representation and indexing of multimedia data are fundamental issues in mobile multimedia technologies, and are becoming an active research direction in the computer society. Traditionally, due to the reliance on low-level feature representation, there is always a tradeoff between accuracy and efficiency. In addition, the overall performance of an underlying model is greatly determined by the distribution of the multimedia data, specially, in a heterogeneous environment. In this chapter, we introduced the content-based representation and indexing of multimedia data in three steps:

1) Four categories of widely-used multimedia-representation approaches (clustering-based, representative-region-based, decision-tree-based, and annotation-based) are discussed in section 2. Their strengths/weaknesses are further compared and listed in a table.

2) The existing content-based indexing models are categorized as two classes: partition-based models and region-based models. In each class, several classic indexing models are outlined. The properties and rationales of these models are compared and contrasted.

3) Considering the deficiency of the existing multimedia representation and indexing schemes, the semantic-based multimedia data manipulation is introduced in section 4. A newly proposed semantic-aware multimedia data representation and organization platform — Extended Summary-Schemas Model (ESSM) — is presented. The ESSM provides a unique logic-based scheme that facilitates the representation and retrieval of multimedia data objects in mobile heterogeneous environments. The ESSM is further contrasted with the other representation and indexing models from various aspects.

Acknowledgement

The National Science Foundation under contract IIS-0324835 in part has supported this work.

References

[1] G. Auffret, J. Foote, C. Li, B. Shahraray, T. Syeda-Mahmood, and H. Zhang, Multimedia access and retrieval (panel session): The state of the art and future directions, *Proceedings of the seventh ACM international conference on multimedia*, 1999, pp 443-445.

[2] I. J. Cox, M. L. Miller, T. P. Minka, T. V. Papathomas, and P. N. Yianilos, The Bayesian image retrieval system, PicHunter: theory, implementation, and psychophysical experiments, *IEEE Transactions on Image Processing*, **9**, 20, (2000)

[3] J. B. Lim, and A. R. Hurson, Transaction processing in mobile, heterogeneous database systems. *IEEE Transaction on Knowledge and Data Engineering*, **14**, 1330, (2002)

[4] W. Hsu, T. S. Chua, and H. K. Pung, Approximating content-based object-level image retrieval, *Multimedia Tools and Applications*, **12**, 59, (2000)

[5] J. B. Kim, and H. J. Kim, Unsupervised moving object segmentation and recognition using clustering and a neural network. *International Conference on neural networks*, 2002, pp 1240-1245.

[6] M. Singh, Segmentation of functional MRI by K-means clustering. *IEEE Transactions on Nuclear Science*, **43**, 2030, (1996)

[7]   B. Li, K. Goh, and E. Y. Chang, Confidence-based dynamic ensemble for image annotation and semantics discovery. *ACM Multimedia*, 2003, pp 195-206.

[8]   B. Heisele, Segmentation of Range and Intensity Image Sequences by Clustering. *International Conference on Information Intelligence and Systems*, 1999, pp 223–225.

[9]   Y. Huang, T. Chang, and C. Huang, A Fuzzy feature clustering with relevance feedback approach to content-based image retrieval. *IEEE Symposium on Virtual Environments*, Human-Computer Interfaces and Measurement Systems, 2003, pp 57-62.

[10] C. Carson, Region-based image querying. *IEEE Workshop on Content-Based Access of Image and Video Libraries*, 1997, pp 42-49.

[11] F. Jing, M. Li, H. Zhang, and B. Zhang, Region-based relevance feedback in image retrieval. *IEEE Symposium on Circuits and Systems*, 2002, pp 26-29.

[12] H. Samet, The Quadtree and related hierarchical data structures. *ACM computing surveys*, **16**, 187, (1984)

[13] J. T. Robinson, Physical Storage Structures: The K-D-B-Tree: A Search Structure for Large Multidimensional Dynamic Indexes. *Proceedings of the ACM SIGMOD International Conference on Management of data*, 1981, pp 10-18.

[14] A. W. Fu, P. M. Chan, Y. Cheung, and Y. S. Moon, Dynamic VP-tree indexing for n-nearest neighbor search given pair-wise distances. *VLDB Journal*, **9**, 154, (2000)

[15] M. R. Rezaee, J. van der Zwet, P. M. E. Lelieveldt, B. P. J. van der Geest, and R. C. Reiber, A Multiresolution Image Segmentation Technique Based on Pyramidal Segmentation and Fuzzy Clustering. *IEEE Transactions on Image Processing*, **9**, 1238, (2000)

[16] B. Heisele, and W. Ritter, Segmentation of Range and Intensity Image Sequences by Clustering. 1999 *International Conference on Information Intelligence and Systems*, 1999, pp 223–225.

[17] D. Yu, and A. Zhang, Clustertree: Integration of Cluster Representation and Nearest Neighbor Search for Image Databases. 2000 *IEEE International Conference on Multimedia and Expo*, 2000, 3, pp 1713-1716.

[18] Y. Konig, and N. Morgan, Supervised and unsupervised clustering of the speaker space for connectionist speech recognition. *IEEE International Conference on Acoustics, Speech, and Signal Processing* (ICASSP-93), 1993, pp 545-548.

[19] J. Wang, W. Yang, and R. Acharya, Color clustering techniques for color-content-based image retrieval from image databases. *IEEE International Conference on Multimedia Computing and Systems*, 1997, pp 442–449.

[20] A. Guttman, R-trees: A dynamic index structure for spatial searching. *ACM SIGMOD International Conference on Management of Data*, 1984, pp 47-57.

[21] N. Katayama, S. Satoh, The SR-tree: An Index Structure for High-dimensional Nearest Neighbor Queries, *Proceedings of the ACM SIGMOD International Conference on Management of Data*, 1997, 26 (2), pp 369-380.

[22] T. Brinkhoff, Efficient Processing of Spatial Joins Using R-trees. *Proceedings of the ACM SIGMOD International Conference on Management of Data*, 1993, 22(2), pp 237-246.

[23] N. Beckmann, H. Kriegel, R. Schneider, and B. Seeger, The R*-tree: An Efficient and Robust Access Method for Points and Rectangles, *Proceedings of the ACM SIGMOD International Conference on Management of Data*, 1990, 19 (2), pp 322-331.

[24] D. A. White, and R. Jain, Similarity indexing with the SS-tree. *Proceeding of the 12th International Conference on Data Engineering.* 1996, pp 516-523.

[25] J. Z. Wang, and J. Li, Learning-based linguistic indexing of pictures with 2-d mhmms. *Proceeding of ACM Multimedia*, 2002, pp 436-445.

[26] M. W. Bright, A. R. Hurson, and S. Pakzad, A Taxonomy and Current Issues in Multidatabase Systems. *IEEE Computer*, **25**, 50, (1992)

[27] M. W. Bright, A. R. Hurson, and S. Pakzad, Automated Resolution of Semantic Heterogeneity in Multidatabases. *ACM Transactions on Database Systems (TODS)*, **19**, 212, (1994)

[28] S. Teller, Application Challenges to Computational Geometry. *Technical Report TR-521-96*, Princeton University, April 1996.

[29] I. T. Young, J. J. Gerbrands, and L. J. van Vliet, *Image Processing Fundamentals*. CRC Press, (2000)

[30] A. Pentland, View-Based and Modular Eigenspaces for Face Recognition, *IEEE Conf. on Computer Vision & Pattern Recognition*, Seattle, WA, July 1994.

[31] M. Li, and D. Li, Dimitrova, N.; Sethi, I. K. Audio-visual talking face detection. *Proceeding of International Conference on Multimedia and Expo*, 2003, pp 473-476.

[32] M. R. Naphade, Detecting semantic concepts using context and audiovisual features. *IEEE Workshop on Detection and Recognition of Events in Video*, 2001, pp 92-98.

[33] D. Li, N. Dimitrova, M. Li, and I. K. Sethi, Multimedia content processing through cross-modal association. *ACM Conference on Multimedia*, 2003, pp 604-611.

[34] D. R. Heisterkamp, Building a latent semantic index of an image database from patterns of relevance feedback. *International Conference on Pattern Recognition*, 2002, pp 134-137.

[35] M. R.Naphade, and T. S. Huang, Recognizing high-level audio-visual concepts using context. *International Conference on Image Processing*, 2001, pp 46-49.

[36] P. L. Lai, and C. Fyfe, Canonical correlation analysis using artificial neural networks. *Proceeding of European Symposium on Artificial Neural Networks* (ESANN), 1998.

[37] J. Hershey, and J. Movellan, Using audio-visual synchrony to locate sounds. *Proceeding of Advances in Neural Information Processing Systems*, 1999, pp 813-819.

[38] S. Chen, and R. L. Kashyap, A spatio-temporal semantic model for multimedia database systems and multimedia information systems. *IEEE Transactions on Knowledge and Data Engineering*, **13**, 607, (2002)

[39] B. Yang, and A. R. Hurson, Similarity-Based Clustering Strategy for Mobile Ad Hoc Multimedia Databases. *International Journal on Mobile Information Systems (IJMIS)*, **1**, 253, (2005)

In: Mobile Multimedia: Communication Engineering ...
Editors: I.K. Ibrahim and D. Taniar pp. 197-215

ISBN: 1-60021-207-7
© 2006 Nova Science Publishers, Inc.

Chapter 10

DATA BROADCASTING SCHEMES INCORPORATED WITH MULTIPLE NON-COLLABORATIVE SERVERS QUERYING

*Say Ying Lim**[*]

Caulfield School of Information Technology, Monash University, Australia

*David Taniar**[**]

Clayton School of Information Technology, Monash University, Australia

Bala Srinivasan

Caulfield School of Information Technology, Monash University, Australia

Abstract

The emergence of mobile computing has opened an increase interest in information dissemination via broadcast in a wireless environment. In this paper, we strive to explore the different ways of data broadcasting mainly periodic broadcasting and on demand broadcasting together with their advantages and disadvantages when encountered in different situations. We also look at some of the issues that arise in a broadcasting environment typically in data scheduling and indexing in multi channel environments. In addition, we will also look at using broadcasting that involves multiple non collaborative servers.

1 Introduction

We have seen an increase demand in which mobile users access information over the wireless channel using their portable devices. This has been growing at an unbelievable rate which is heavily reinforce and encouraged by the rapid development as well as recent advances in wireless network technologies that have led to the development of the concept mobile computing [5,27]. Mobile computing has provides mobile users the ability to access

---

[*] E-mail address: Say.Ying.Lim@infotech.monash.edu.au
[**] E-mail address: David.Taniar@infotech.monash.edu.au

information anytime, anywhere. It enables mobile users to query databases from their mobile devices over the wireless communication channels [15]. It differs from the conventional computing due to the mobility of nomadic uses and their portable devices as well as their inherent environment and resources limitations such as low storage, asymmetric communication cost, bandwidth limitation as well as limited battery life [5]. Despite all these limitations, mobile computing is still becoming more and more in demand.

Lately, there has been an increasing interest among researchers in the area of mobile computing to look at information systems that deliver data using broadcasting in a wireless environment. This strategy is also known as the broadcasting data delivery in which a server perform data broadcast to the users through wireless channels [3,14]. As the number of users can be extensively large, it is important to go for techniques that are able to produce low response time to data request. This operation differs from the conventional pull-based environment whereby servers deliver data by separately responding to individual request query that are sent by the users. Generally, there are two basic ways to for data broadcasting which are periodic broadcasting (push based broadcasting system) and on demand broadcasting (pull based broadcasting system)

Example 1: A broadcast system can be illustrated as such situation whereby in the airport, schedules that are up to date about the flight details are being broadcast to clients that are within the coverage. This would enable mobile users with their mobile devices be able to receive the updated schedules.

Broadcasting or also known as the 'On-Air' strategy appears to be one type of query processing methods besides the server and client strategies. It has been proven to be an effective method in coping with massive number of mobile clients. Sometimes, when obtaining information from a single server is deemed to be not enough and mobile users may therefore need to obtain from several servers to be combined [24,25]. And so by studying on how to gather this different information from the remote non-collaborative servers together is significant in the user's point of view. Actually, the work presented in this paper is part of a larger project whereby some of the results have been previously published [24,25]. However, in the earlier work, we consider multiple non-collaborative servers that only accepts direct query and is utilizing a server strategy and does not concern broadcasting.

When encountering the need to query data from multiple non-collaborative servers, we are actually referring to multiple individual servers that do not hold any relation between one another. They play their own individual functions with no respect to each other. Their purpose is just to distribute their data to the mobile users that are within the active region. Hence, non-collaborative servers are independent service providers.

In this chapter, we begin with providing some awareness to the introductory of a mobile environment and some knowledge on non collaborative servers and queries which can all be found in Section 2. Then in Section 3, we look at broadcasting in greater detail together with their advantages and disadvantages followed by Section 4 where we explore on a more technical aspect which include the issues of broadcasting, such as scheduling and indexing as well as the use of multi channels. In Section 5, we describe some algorithms for information gathering involving multiple non-collaborative servers querying that involves data broadcasting. Lastly is Section 6 that concludes the paper together with a plan for the future in Section 7.

2   Background

The effect of having the ability to broadcast a number of data items to a large number of users simultaneously is of great importance especially in the mobile computing [5,27]. This is due to the reasons that contacting the remote servers individually for data is more expensive and slow due to mobile users need to queue for their turn to request for data items in the wireless environment and there is a limited dedicated point to point connection that can be provided at one particular time. In this section, we first introduce the mobile computing environment, followed by a general idea of non-collaborative servers and mobile query processing.

2.1   Background

Generally, mobile users with their mobile devices and servers that store data are involved in a typical mobile environment [5]. Each of these mobile users communicates with a single server or multiple servers that may or may not be collaborative with one another. Typically, each mobile user communicates with the server in order to carry out any transaction and information retrieval. Basically the servers are more or less static and do not move, whereas the mobile users can move from one place to another and are therefore dynamic. Each of these servers communicates through a wireless interface with the mobile clients and it servers a large amount of clients within certain coverage in a specific region. Thus, mobile users have to be within specific region to be able to received signal in order to connect to the servers [18].

Figure 1 illustrates a scenario of a mobile database environment which involves mobile users moving from one location to another location. Whenever mobile users are within a specific region or cell, he/she can access to information that are provided by the servers within that region. And the mobile users will obtain the desired data and downloaded into their mobile device. And when they move to a different location, the information that they access may have been change due to the changing of the region that has different server providing different information. So again when in a new region, the mobile user can download the desired data and with this two individual piece of data that are obtained from the different servers in different location, the mobile device can manipulate the data locally. Assuming mobile client $C3$ currently in Location B and is accessing server $S2$, once he got the desired data, he moves to Location C and is now accessing server $S3$ which provide different data and vice versa. Hence, the change of data obtained that are reflected by the movement of the mobile users has introduces the concept of location dependent queries [28].

Example 2: A property investor while driving his car downloads a list of nearby apartments for sale from a real-estate agent. As he moves, he downloads the requested information again from the same real-estate agent. Because his position has changed since he first enquires, the two lists of apartments for sale would be different due to the relative location when this investor was inquiring the information. Based on these two lists, the investor would probably like to perform an operation on his mobile device to show only those apartments exist in the latest list, and not in the first list.

Thus, in a classic mobile environment, it is common to meet with situations where a mobile user who is currently in an active region obtaining data that he/she is not satisfied with and may require further processing with other data that can only be obtained from other servers that may be location within the same region or maybe from another different region. This shows the need to do query processing involving multiple servers that may not be collaborative between each other.

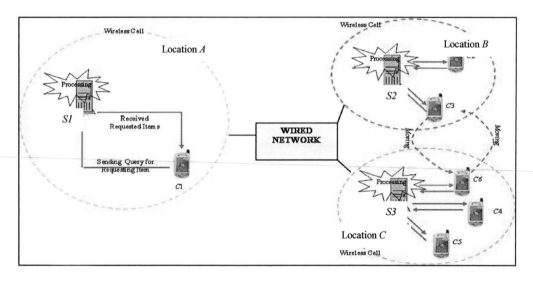

Figure 1: A Mobile Database Environment

2.2   Non-Collaborative Servers and Queries

Within a certain cell or region, there are multiple servers around and they can be collaborative or non-collaborative. However, the focus in this paper, we will only look at servers that are non-collaborative. Basically, non-collaborative servers would refer to servers that do not know each other and do not have any relation between one another. There are basically just individual server providers which disseminate data to the users and they do not communicate with one another [24,25]. Each of these servers serves as an independent service provider, and often these independent servers are only specialized within the domain the information they are providing.

Furthermore, the servers can have their ways to disseminate their information to the mobile users within the cell. Some servers may only broadcast data and may not accept queries whereas some servers can accept queries and then broadcast data [39]. And with the variation of the ways that a server disseminates their information, several algorithms will be studied in this paper.

Example 3: A mobile user may want to know what the timetable for the transportation services to a particular event. Each of the transportation timetable as well as the event is stored in different servers, which is transport server and event server. Transport server would deal with transportation data while an event server would deal with current events that are happening. Therefore in order to know transportation timetable for a particular even, the user have to gather data from the two different servers which are non- collaborative.

In addition, not all service providers are supported by the usage of a mediator. Therefore information obtained from other independent non related service providers need to be process individually. It is not a fair assumption that every service providers are linked through a mediator. Hence, in our research we focus independent service providers which are referring non collaborative servers. Thus, it is vital to consider gathering information from non collaborative servers because it is often not enough to just get data from a single server.

From the query processing perspective, the most important element is to help reduce the communication cost, which occurs due to data transfer between to and from the servers and mobile devices [23]. The need for collecting information from multiple remote databases and processing locally becomes apparent especially when mobile users collect information from several non-collaborative remote databases.

3   Data Broadcasting Systems

In general, there are two basic methods of data broadcasting that can be adapted. The most straight forward method would be purely broadcast the data without giving any control to the mobile users as in multiple non collaborative servers setting, some servers may not able to accept queries that are directly issued by the mobile users. This is known as *periodic broadcasting* (push based broadcasting system [1,2,6,7,29,34].

The second method to broadcast data item would be known as *on-demand broadcasting* (pull based broadcasting system). This way of data broadcasting is almost similar to the server strategy where mobile users is given the ability to initiate the data transfers first by sending a query request to the server [2,4,32].

3.1   Periodic Broadcasting (Push Based Broadcasting System)

This method refers to the server periodically broadcasts data items to an arbitrary number of mobile users within the active region that allows any mobile users who is interested in the data that are being broadcast to tune into the broadcast channel to select the desired item [34,36,37,38]. If this is the situation, then the users would not be able to obtain the desired information immediately but depending on where the desired data item is allocated. This server may serve as a broadcasting system, which can only broadcasts data to the air and the user have to tune to the air in order to obtain the desired data [20]. This type of system will therefore broadcast data items within the scheduled time regardless of whether that data item is demanded or not at the particular time.

Example 4: For instance, a mobile user only wants to find out sports shop during his visit to a shopping complex. He then issues a query to obtain his information and the response to this query depends on the location of the mobile user. The data broadcasting supported by the shopping complex is push based type system and will broadcast information of all the shops that are available in the shopping complex, then the mobile user would have to tune and look at which is the sports shop and then download it.

As in Example 4, the mobile user has to wait until his desired sports shop information to arrive as the server does not accept direct query and will therefore broadcast every single data items. The user would tunes into the broadcast channel and manually filters out the relevant desired data. The user has no control in issuing queries directly to the servers and that is why the server will broadcast all the shops information that is available to all users within the cell region. This type of broadcasting system allows an arbitrary number of users accessing the data simultaneously. However the drawback of this is crucial as it may be acquainted with issue of over population in accessing a particular data that may slow down the access.

Figure 2 models a periodic broadcasting system where mobile users obtain the desired data items through tuning into the broadcast channel and capturing the data whenever it arrives.

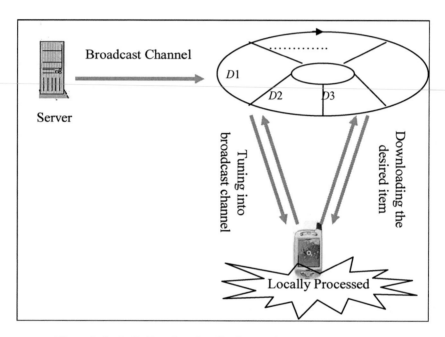

Figure 2: Periodic Broadcasting (Push Based Broadcasting System)

3.2   On-Demand Broadcasting (Pull Based Broadcasting System)

The on-demand broadcasting method allows mobile users to have more control and flexibility in comparison to the periodic broadcasting system since in this method, the server would only broadcast data items that are requested by the mobile users. Then the mobile users would queue up in the same order as the requests order they have made. The server will make a schedule to satisfy the mobile user's request. With this method, the user does not have to wait for a substantial amount of time for the arrival of his data item from the broadcast channel [1,2].

Example 5: Similar to Example 4, with the difference that instead of broadcasting all shop information in the shopping complex, the mobile user has the control of sending a direct query to request for only Sport shop information. And with adopting an on demand system, the user can send direct query to obtain only Sports Shop information and the server will therefore only broadcast the desired data item. Figure 3 models an on demand broadcasting where it only broadcasts the desired data items to the available mobile users.

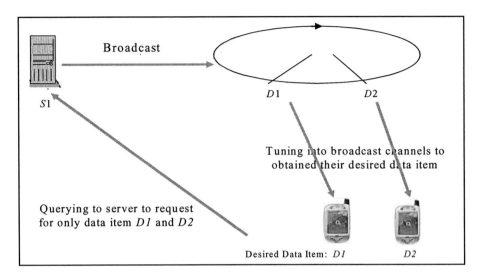

Figure 3: On Demand Broadcasting (Pull Based Broadcasting System)

3.3   Advantages and Disadvantages of Data Broadcasting

Regardless of whether the query are being processed on the server of on the client side, and whether the user has to tune into the channel to obtain the desired data items or to send a direct query to the server, there are both advantages and disadvantages in using the strategy in a mobile environment in certain circumstances. This section will explore on the pros and cons of data broadcasting in general despite of whether it is a on demand or periodic broadcasting.

3.3.1   Advantages of Data Broadcasting

The most inherent advantage of data broadcasting is due to its effective and efficient way to make data accessible to a large number of populations. This is because of its well known excellent scalability [16,17]. This has made scalability especially important due to the intrinsic nature of the mobile environment which has limited wireless resources [26,40].

As far as cost remains a major concern to a wide majority of mobile users, data broadcasting is believe to be more cost effective and appear to be more economically viable than establishing a direct communication to the server which is as how server strategy providing exclusive point to point communication between a certain user and server which in this case the server process the query that are being sent by the mobile users and return back the results to the mobile users [32]. This will severely affects the query response rate as well as power consumption of mobile users as it involves queuing to sending the request to the server and cannot scale over the capacity of the server.

Thus, a data broadcasting access mode basically appears to be advantaged in a wireless environment since it consumes less bandwidth and emerge as an energy efficient way for mobile users to obtain desired data items [17,29]. With the remarkable feature of data broadcasting in contrast to server responding to individual request, data broadcasting provide a higher throughput for data access to a large number of users [6]. By eliminating communication contention among the mobile users requesting data allows the mobile users to share the available bandwidth. Hence, this can be evaluated in terms of bandwidth available and price rate. As a matter of fact, energy efficient can be gained from this access mode because the mobile users it able to retrieve information without wasting power to transmit request to the servers. This avoids energy expensive request sending.

Also, data broadcasting is distributed in nature which means it provides resources to a number of clients in a fair manner and mobile users will get a same chance to tune into and obtain the desired data [7]. Any mobile users, who are receiving the desired data items, will not affect the performance of other users who are obtaining or monitoring the same data.

3.3.2   Disadvantages of Data Broadcasting

Although the broadcasting appear to be more scalable in comparison to the server strategy, but there is still limitations that it brings to the users because most use would find it easier to send a direct request to a specific server rather than having to tune into the air to search for their desired data that are being broadcast by multiple non- collaborative servers. Therefore, it is crucial to accommodate the request of the user to bring it more flexibility like how server strategy does for the user.

The main challenges would be able to minimize query response time as well as the tuning time in retrieving the required database items [4] It is known that as the data size increases the required number of data items to be broadcast also increases respectively which may in return causes the mobile users having to wait for a reasonably amount of time before being able to received the desired data.

Especially for periodic broadcasting, mobile users has to rely solely on the scheduling of the data items on the broadcast channel to obtain its desired item unlike in a pull-based environment where mobile users plays a more active role in getting the desired data items [2,32]. By having the mobile user to rely solely on monitoring the broadcast channel to obtain the desired data items can turn out to be a disaster especially when the server also perform pull based request and contain a series of continuous request from mobile users. This can easily caused a scalability bottleneck with a large mobile user population.

On-demand broadcasting may cause increase in exceeding usage of bandwidth especially when too many data requests are being sent out by the mobile users. With the overwhelmed mobile users request may affect the query performance. Hence, periodic broadcasting are sometimes useful in certain situations to avoid possible congested channel bandwidth and long server queue which can be cause by the increase number of mobile users.

Another potential drawback especially in a data broadcasting is that active listening on the wireless channel is required as the mobile users has to wait for the required data items to appear in the broadcast channel and this cause energy consumption and need to be taken into consideration since battery life is limited in a mobile device. Several techniques have been proposed that alleviate these drawbacks by the use pre-fetching [1].

4   Data Broadcasting Issues

When determining the design of a broadcasting systems in which the number of mobile users are not dependent on the dedicated point to point lines being provided by the server, several issues need to be taken into account [22,30]. One of the important issues is to determine an optimal broadcast sequence for the data items that are to be distributed to the mobile users. This refers to data broadcast scheduling which is an important factor that one must look at in order to have minimal access time and minimal tuning time in receiving the required data items [4,32,40]. It is important to have a balance between users expectation versus system performance [2].

Besides, concentrating on the order of the data whether which data items should go first and which is next, there are other issues that must be considered, such as, in order to have an efficient data retrieval scheme in broadcasting, factors such as the use of indexing and multi channels can aid in providing efficient data and this to assist mobile users to perform query operation more efficiently [10,36-38].

4.1   Data Broadcast Scheduling

Data broadcast scheduling is one of the main challenges in data broadcasting which require minimal access time as well as tuning time in receiving the desired data items. Access time indicates the time between sending the query and receiving the results back as response. The access time determines the response time of query made by mobile users and it is vital to make the average access time smaller. As for tuning time, it indicates the total amount of time the mobile users spend actively listening to the broadcast channel to obtain the desired data items. Tuning time determines the power consumption of mobile users. It is crucial to have reduction in tuning time because most mobile users depend on limited battery life on their devices [1].

Query response time can be reduced by having a good selection mechanism. One can a prioritized broadcast sequence especially important due to mobile device at the client side have significant constraints in terms of energy, storage and computation power. As for distributing the broadcast data over more than one broadcast channel, it is believe to improve the response time as well as organizing data items over the broadcast channel [8,22]. There have also been ways proposed in existing work dealing with tuning time problem which include inserting index segments into broadcast channel [17]. By having incorporating application semantics into indexing technique can help achieve better performance as well [20].

By prioritizing the data items and using a good selection mechanism can reduce the broadcast cycle length which eventually is able to reduce the query response time [11,12]. The data items can be characterized as both 'Hot' and 'Cold' and this can be the determinant of whether which data item should be given a higher priority over the others.

Example 6: A traveler driving along a highway is more likely to stop over for a petrol kiosk than for a video shop. And with this assumption, along the highway there will be more petrol kiosks than video shops. So the traveler would more likely to make query for information regarding the petrol kiosk and these data items are therefore access more frequently than video shop. Hence, the more frequently accessed data is known as 'Hot' data items and should be given higher priority when determining which data items should be placed in the broadcast channel.

The other factor that could help reduce response time would be concerning the organization of the data items especially when retrieving multiple data items is required [8,22,38]. An illustration of such a situation can be a mobile client want to send a query to retrieve multiple stock prices concurrently. This is an example of multiple data items retrieval and in order to retrieve such query in a more efficient way, the need to consider the semantic relationship between the data items is required. With a proper organization of the data items over the broadcast channel can reduce the access and downloading time. This is because by allocating related data items in a manner that may appear in the way the mobile client would want to retrieve in a sequence order, the mobile client does not have to wait too long for the next related data item after he/she has download a particular data item. Hence, organization of the data item is designed in a way that matches the possible mobile client query access pattern [8,10].

However, in order to predict which data item that the mobile client would be interested next is difficult because there is not much knowledge of any future query that is being available. Existing related work has investigated on the use of access graph to represent the dependency of the data items [22]. Other existing algorithms that has been investigated on to identify the most effective organization of the data items includes heuristics algorithm [14], and randomized algorithm [6]. Besides these, there are many other algorithms that serve the same purpose.

4.2   Indexing and Multi Channels Broadcasting Systems

The last possible deciding factor for query response time can be determined by incorporating broadcast indexing scheme. The use of indexing in the broadcasting environment is needed to inform the mobile users when their desired data item will arrive in the broadcast channel. Indexing scheme can reduce tuning time for the mobile client to access their required data item [20].

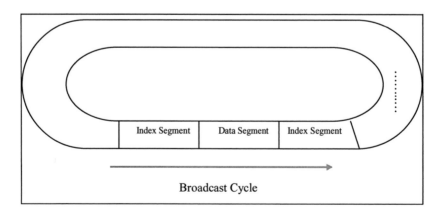

Figure 4: Index Scheduling

By applying this scheme, mobile clients can conserve their battery life and thus result in energy saving because the clients can switch to "doze" mode and back to "active" mode only when he/she knows the desired data item is about to arrive. The client is able to know this is because with the adoption of indexing, some form of additional data will be broadcast together with the data items and the client can first obtain the index to know when the required data item is about to arrive since the additional index information will contain which data items are in order in the broadcast channel together with the exact time of the data to be broadcast. However the results of this usually require a trade-off between reducing tuning time and reducing query response time. Hence, it is important that when utilizing indexing scheme is to choose the optimal trade-off balance the tuning and response time. Figure 4 illustrate an example of index scheduling whereby the index is place together with the data broadcast.

Some of the existing work that researcher have been work on that could be used in a broadcast environment include round-robin index, random index, and windowed indexing. These techniques are use to broadcast indexing information for a given set of data items. Round-robin indexing is the most straight forward and naïve method whereby the index information is send together one by one with each file [19,21]. This looks as if it consumes large amount or power and is regard as energy inefficient.

On the other hand, random index will look at which data item are "hot" and therefore in this scheme, it transmit index entries base on the popularity of the file and they are transmitted more often so that it can be located more quickly since its in more demand. The downside of this method is that mobile clients who want to tune to "cold" data item would have to wait for a substantial amount of time for the index information to be arrived. Windowed index is a better approach where it eliminated the infinite waiting time for "cold" data item. A fixed window size is defined and index is sent multiple times each time within a particular defined window size.

Another alternative solution to speed up data retrieval that takes into account reducing response time is to have more than one broadcast channel whereby the broadcast data can be distributed to more than one broadcast channel. In most cases, data items are broadcast over single channel as it avoid additional issue of the organization of data and allocation while having more than one channel [9,16,38,30].

Furthermore, the use of a single channel appears to be more problematic especially when there are large number of data items to be broadcast Thus, with the adoption of multiple channels to broadcast data, the chance of reducing long delays before obtaining the desired data items can be achieved as well as helping to solve problems such as data distortion and data distribution [23]. The reduction time of accessing and having the server to response to the query can be achieved with the usage of multiple channels in comparison to utilizing a single channel to broadcast the data.

Figure 5 exemplify the big picture of multi broadcast channels architecture followed by Figure 6 which shows a more detail breakdown on achieving the multiple channels architecture where we can see that the original channel is broken down into two channels (e.g. Channel #1 and Channel #2) with each of the new channels holding three data items. Assuming in this case is that the original broadcast channel contains six data items that can be split into two small broadcast channels to provide efficient data retrieval.

Example 7: In the highway, there are 10 nearby petrol kiosk. A traveler sends query for the nearby petrol kiosk and he/she only wants to know the Shell Brand petrol kiosk. The Shell Brand petrol kiosk is so happen to be allocated on the 8th data item of the broadcast channel. And if a single channel is being use, then the traveler have to wait for the 8th data item which is his/her desired data item. However, if 2 broadcast channels are being used where by the 1st

channel consist of the first 5 data items and the 2nd channel consists of the last 5 data items, then the traveler only need to wait for the 3rd data item of the 2nd channel.

Thus, in Example 7, the time of accessing and the response time of the query is greatly reduced since the traveler need only wait for the 3rd data item of the 2nd channel instead of the 8th data item when only a single channel is being used. This has show that multiple channels appear to be a more efficient data access method.

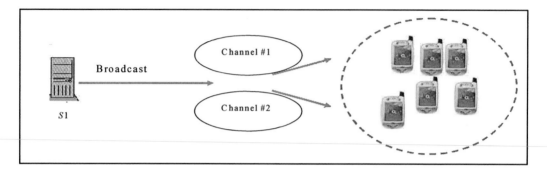

Figure 5: Big Picture of Multi Channels Architecture

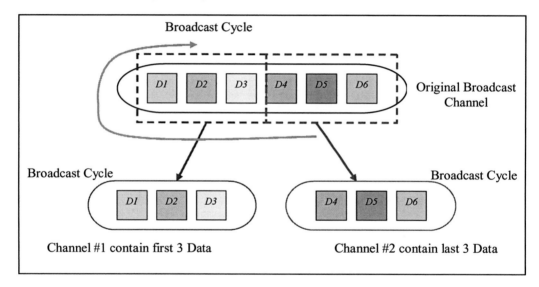

Figure 6: Detail Picture of Multi Channels Architecture

5   Incorporating Broadcasting in Non-collaborative Servers Querying

As described in the earlier sections, that often in a mobile environment not all servers are able to accept direct mobile users queries [31,33,35]. There will be situations when some servers would only able to broadcast data regardless of whether that is the requested item or not whereas others may only accept query and be process in the server or another case is retrieving some part of data from local memory and process them locally. Sometimes it may be useful to gather information from several servers to be process into one useful piece of information.

Figure 7 illustrates an example of a server that supports data broadcast in conjunction with another server that supports direct query. Thus, in order to gather data from the two non-collaborative servers whereby one server support data broadcast and another server support direct query, the mobile users would need to tune into the wireless channel to obtain the desired data from the server that support data broadcast and issue a query to the other server which accept direct query. And with these two pieces of information obtained from servers utilizing different strategy the mobile user can process into a single valuable information locally on the mobile device.

As a result of the importance of information gathering, in this section we would like to study several algorithms that involve information gathering from several non-collaborative servers that uses different ways to process their information. This includes server that accept purely broadcast which we refer to "On-Air Strategy/Broadcast Strategy" and the other server which needs to obtain user request before being process which we refer to "Server Strategy".

We divide the propose techniques into two parts, namely: mobile device side processing and server side processing. Each of these techniques has a different impact in return to the transfer cost and transfer completion time which in certain condition, one can outdo the other depending on the data they wish to obtain and where they currently located.

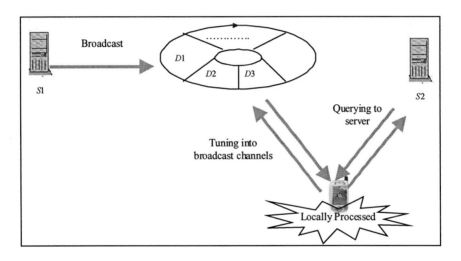

Figure 7: Information Gathering involving Broadcast Stratgy and Server Strategy

5.1   Mobile Device Side Processing (MDSP)

In general, mobile device side processing deals with acquiring information from the servers to the mobile device to be processed locally on the mobile device. Regardless of which type of strategy the server adopt, all processing will be done on the mobile device itself once all the information are obtained individually from the respective servers.

Example 8: For instance a traveler wants to know which nearby Chinese Restaurants is reachable by using a Tram and the timetable of the Tram. In order to know the Chinese

Restaurant the traveler need to communicate with the Restaurant server and in order to get the Tram timetable the traveler needs to talk to the Transportation server. Assuming that the Restaurant server uses server strategy to communicate with the mobile users and Transportation server uses the broadcast strategy and the mobile user want to know the tram timetable to the Chinese Restaurant. MDSP Steps involves:

a) Send a query to the Restaurant server to obtain nearby Chinese Restaurants
b) The server will process the query and send back the results to the mobile user indicating all the nearby Chinese Restaurants
c) With the list of Chinese restaurants on hand, it will first be kept in the memory of the mobile device to be process later on.
d) Next, the mobile user tune into the broadcast channel that are currently broadcasting the timetable of the available transportation route. Since the server is not being able to accept any direct request, the user has to wait until the desire data item to arrive before he/she can download the information
e) Once the desired data item arrive through the broadcast channel, the mobile user can download them into the mobile device. In this case is the Tram information.
f) Now that, the mobile user have successfully obtained the two different lists of information, the mobile device can do a local processing to see if there is any tram that are going to any of the nearby Chinese Restaurants and display the timetable and the Chinese restaurants on the screen.

This method appears to be the most naïve and straightforward method but have several drawbacks. The main drawback is the finite capacity of the memory and therefore it is not realistic to store everything into the device to be process later on. The memory consumption to store everything is way too high and may not be able to store everything.

5.2   Servers Side Processing (SSP)

Due to the possible large amount of data to be downloaded from both servers, the mobile device may not be able support due to the limited memory and processing capability. This section investigates other alternative, such as Server Side Processing (SSP). Figure 8 models an example of the architecture of SSP. Using the same Example 8 below are the steps that involves in SSP:

a) The mobile user first tune into the broadcast channel that are broadcasted from Server 1 (S1) which are currently broadcasting the timetable of the available transportation route by waiting until the desire data item to arrive before he/she can download the information
b) Once the desired data item which is the available tram information arrives through the broadcast channel from S1, the mobile user can download them into the mobile device.
c) With the list obtained via the Broadcast strategy on hand which is the Tram information, send the list to the Restaurant server which is S2 to be process and

obtain the matching information base on the Tram destination with the restaurants location.

d) The server will process the query and send back the results to the mobile user indicating all the nearby Chinese Restaurants which has trams that are going there.

e) Now that we know which tram are going to which nearby Chinese restaurant, the user can now tune into the broadcast channel again and retrieve the additional information regarding the timetable of that particular tram that are going to the user desired Chinese restaurant.

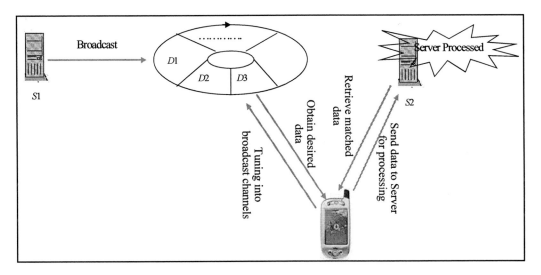

Figure 8: Information Gathering involving Broadcast Stratgy and Server Strategy

This method has eliminated the need to keep one full list of information and therefore consume less memory. Also less power consumption because part of the processing power has been done by the server which performs the matching. Although more round trips is involve and having the user to tune into the broadcast channel for the second time can be a drawback, but the limitation of mobile device memory can be overcome.

6   Conclusions

Rapid advances in wireless technology and potable information devices have increased the popularity of mobile computing which is a new paradigm of computing. There are still several limitations that are inherent be it the mobile devices it self or the environment itself. These include limited battery power, storage, communication cost, and bandwidth problem. For this reason, broadcasting appears to be a general method to deliver data to mobile users in mobile computing applications especially with its ability to reduce energy expenditure and scalability feature.

In this paper, we started off with introducing data broadcasting in a mobile environment and further explored the pros and cons of the two basic ways of data broadcasting which is (1) periodic broadcasting and (2) on demand broadcasting. This paper also looks at how we can gather information using the different strategies that each individual non-collaborative servers are adopting to obtain one valuable piece of information since often a case where obtaining information from just one server is not enough and also sometimes arises situations where multiple servers are adapting a different strategies when comes to disseminating data.

7   Future Work

There have been several researches done in the area of exploring data broadcasting in a mobile environment. There are still many limitations of the nature of the mobile environment as well as mobility of the users that generates a lot of attention from researching in finding a good broadcast strategy that can cope well with frequent disconnection and low power consumption and that is able to minimize the tuning and query response time. Investigating on the anticipation regarding the knowledge of related future query that mobile clients might make is also beneficial so that the data in the broadcast channel can be arrange in a better order to reduce response time.

As a part of our future work, it is critical to design algorithm that involves gathering information using other strategies such as client strategy. We will also look at obtaining partial information from other clients through their cache memory and gather them with their user own cache or obtain new information from the server. In addition, in the future work, we plan to create more techniques and algorithms and evaluate them by producing cost models and make extensive comparisons. Having exploring to a certain degree of data replication would probably help [13].

Another related area that is of great interest to be pursues is caching which could cope with the frequent disconnection and frequent updates that may occur when accessing data in the non collaborative server as well as dealing with situations when the user wishes intersect with data that are being downloaded from two non collaborative servers and to know which is to cache and which is not.

We plan to extend our current work to deal with location based broadcast information services where geographical coordinates comes into place. Thus, we intend to propose algorithm that is able to decrease the access time and to also have the organizing time short as well as minimizing movements of users. This refer to having our propose algorithm that can helps reduce the mobile user having to go back to the previous location to obtain additional data if he forgets when he was there before. This is useful especially when the user is require to obtain data from several non-collaborative servers that are not within a single region but are spread into several different regions. This can appear to be challenging because at the same time, it is important to make sure memory consumption to keep the data in the local memory remain low.

References

[1] Acharya S., Franklin M. and Zdonik S., "Prefetching From A Broadcast Disk", *Proceedings of 12th International Conference on Data Engineering,* pp.276-285, 1996.
[2] Acharya S., S. Muthukrishnan, "Scheduling On-Demand Broadcasts: New Metrics and Algorithms", *MOBICOM 98,* pp.43-54, 1998.

[3] Aksoy D., Altinel M., Bose R., Cetintemel U., Franklin M., Wang J., and Zdonik S., "Research in Data Broadcast and Dissemination", *Proceedings of 1st International Conference on Advanced Multimedia Content Processing, LNCS* **1554**:194-207, 1999.

[4] Aksoy D., and Franklin M., "R X W: A Scheduling Approach for Large Scale On-Demand Data Broadcast", *IEEE/ACM Transaction On Networking*, 7(6):846-860, 1999.

[5] Barbara D.,"Mobile Computing and Databases-A Survey", *IEEE Transactions on Knowledge and Data Engineering,* 11(1):108-117, 1999.

[6] Bar-Noy, A., Naor, J., and Schieber, B., "Pushing Dependent Data in Clients-Providers-Servers Systems", *Proceedings of the 6th ACM/IEEE on Mobile Computing and Networking,* pp. 222- 230, 2000.

[7] Bar-Noy A., and Shilo, Y., "Optimal Broadcasting of Two-Files over an Asymmetric Channel", *Journal of Parallel and Distributed Computing*, 60(4):474-493, 2000.

[8] Chung Y.D. and Kim M.H., "Effective Data Placement for Wireless Broadcast", *Distributed and Parallel Databases*, 9(2):133–150, 2001.

[9] Hu Q., Lee D. L. and Lee W.C., "Optimal Channel Allocation for Data Dissemination in Mobile Computing Environments", *Proceedings of 18th International Conference on Distributed Computing Systems*, pp.480-487, 1998.

[10] Hu Q., Lee W.C. and Lee D. L., "Indexing Techniques for Wireless Data Broadcast under Data Clustering and Scheduling", *Proceedings of the 8th ACM International Conference on Information and Knowledge Management*, pp.351-358 1999.

[11] Huang J.L. and Chen M.-S., "Dependent Data Broadcasting for Unordered Queries in a Multiple Channel Mobile Environment", *Proceedings of the IEEE GLOBECOM*, pp.972-976, 2002.

[12] Huang J.L. and Chen M.-S., "Broadcast Program Generation for Unordered Queries with Data Replication", *Proceedings of the 8th ACM Symposium on Applied Computing (SAC-03)*, pp.866-870, 2003.

[13] Huang Y., Sistla P. and Wolfson O., "Data Replication for Mobile Computers", *Proceedings of the ACM SIGMOD,* pp.13-24, 1994

[14] Hurson A.R., and Jiao Y., "Data Broadcasting in Mobile Environment", *Wireless Information Highways,* Katsaros D., Nanopoulos A., and Manolopoulos Y. (editors), Chapter 4, IRM Press Publisher, London, 2005.

[15] Imielinski T. and Badrinath B. "Mobile Wireless Computing: Challenges in Data Management", *Communications of the ACM,* 37(10):18-28, 1994

[16] Imielinski T., Viswanathan S. and Badrinath B. R., "Energy Efficient Indexing on Air", *Proceedings of the ACM SIGMOD Conference*, pp.25-36, 1994.

[17] Imielinski T., Viswanathan S. and Badrinath B. R., "Data on Air: Organisation and Access", *IEEE Transactions on Knowledge and Data Engineering,* 9(3):353-371, 1997.

[18] Jayaputera, J. and Taniar, D., "Location-Dependent Query Results Retrieval in a Multi-cell Wireless Environment", *Parallel and Distributed Processing and Applications, Lecture Notes in Computer Science*, Vol. 3358, Springer-Verlag, pp. 49-53, 2004

[19] Khanne, S., and Liberatore, V., "On Broadcast Disk Paging", *SIAM Journal on Computing,* 29(5):1683-1702, 2000.

[20] Lee C. K. K., Leong H. V. and Si A., "Adaptive Semantic Data Broadcast in a Mobile Environment", *Proceedings of the 2001 ACM Symposium on Applied Computing*, pp.393-400, 2001.

[21] Lee D. L., Hu Q. and Lee W.C., "Indexing Techniques for Data Broadcast on Wireless Channels", *Proceedings of the 5th Foundations of Data Organization*, pp. 175-182, 1998.

[22] Lee G., Lo S-C., and Chen A.L.P., "Data Allocation on Wireless Broadcast Channels for Efficient Query Processing", *IEEE Transactions on Computers*, **51**(10):1237-1252, 2002.

[23] Leong H. V. and Si A., "Data Broadcasting Strategies over Multiple Unreliable Wireless Channels", *Proc of the 4th Inform & Knowledge Manag*, pp.96-104, 1995.

[24] Lim, S.Y., Taniar, D., and Srinivasan, B., "On-Mobile Query Processing incorporating Multiple Non-Collaborative Servers", *ISI Journal, Special Issue on Mobility in the Information Systems and Databases*, 2005, **10**(5), 2005

[25] Lim, S.Y., Taniar, D., and Srinivasan, B., "On-Mobile Aggregate Query Processing incorporating Multiple Non-Collaborative Servers", *Proceedings of the International Conference on Advances in Mobile Multimedia MoMM 2005*, pp. 21 – 30, 2005.

[26] Malladi R. and Davis K.C., "Applying Multiple Query Optimization in Mobile Databases", *In Proceedings of the 36th Hawaii International Conference on System Sciences*, pp.294-303, 2002.

[27] Myers B.A., and Beigl M., "Handheld Computing", *IEEE Computer Magazine*, **36**(9):27-29, 2003.

[28] Park K., Song M., and Hwang C-S, "An Efficient Data Dissemination Schemes for Location Dependent Information Services", *Proc of the 1st Intl Conf on Distributed Comp &Internet Tech (ICDCIT 2004)*, Vol. 3347, Springer-Verlag, pp.96-105, 2004.

[29] Papadopouli, M and Issarny, V., "Effects of Power Conservation, Wireless Coverage and Cooperation on Data Dissemination among Mobile Devices" *Proc. of the 2nd ACM Intl. Symp.on Mobile AdHoc Networking & Computing (MobiHoc)*, pp.117-127, 2001.

[30] Prabhajara, K., Hua, K.A., and Oh, J.H., "Multi-Level, Multi-Channel air Cache Designs for Broadcasting in a Mobile Environment", *Proceedings of the 16th International Conference on Data Engineering*, pages 167-186, 2000.

[31] Si A. and Leong H. V., "Query Optimization for Broadcast Database", *Data and Knowledge Engineering*, **29**(3): 351-380, 1999.

[32] Sun W., Shi W. and Shi B., "A Cost-Efficient Scheduling Algorithm of On-Demand Broadcasts", *Wireless Network*, **9**(3):239-247, 2003

[33] Su, C.J., Tassiulas, L., and Tsotras, V.J., "Broadcast Scheduling for Information Distribution", *Wireless Neworks*, Vol.5, pp. 137-147, 1999.

[34] Triantafillou P., Harpantidou R. and Paterakis M., "High Performance Data Broadcasting: A Comprehensive Systems 'Perspective", *Proceedings of the 2nd International Conference on Mobile Data Management (MDM 2001)*, pp.79-90, 2001.

[35] Waluyo, A.B., Srinivasan, B., and Taniar, D., "A Taxonomy of Broadcast Indexing Schemes for Multi Channel Data Dissemination in Mobile Databases", *Proceedings of the 18th International Conference on Advanced Information Networking and Applications (AINA 2004)*, Vol. 1, IEEE Computer Society Press, pp. 213-218, 2004.

[36] Waluyo, A.B., Srinivasan, B., and Taniar, D., "Indexing Schemes for Multi Channel Data Broadcasting in Mobile Databases", *International Journal of Wireless and Mobile Computing*, **1**(6), 2005.

[37] Waluyo, A.B., Srinivasan, B., Taniar, D., and Rahayu, J.W., "Incorporating Global Index with Data Placement Scheme for Multi Channels Mobile Broadcast Environment",

Embedded and Ubiquitous Computing, Lecture Notes in Computer Science, Volume 3824, pp. 755-764, Springer-Verlag, 2005.

[38] Waluyo, A.B., Goh, G., Srinivasan, B., and Taniar, D., "Applying Multi-Channelling and Indexing Scheme for Efficient Data Retrieval in Wireless Broadcast Environment", *Proceedings of the Second International Conference on Intelligent Sensing and Information Processing (ICISIP'05)*, IEEE Computer Society Press, pp. 32-37, 2005.

[39] Wong J.W., "Broadcast Delivery", *Proceedings of the IEEE*, 76(12):1566-1577, 1998.

[40] Yajima E., Hara T., Tsukamoto M. and Nishio S., "Scheduling and Caching Strategies for Correlated Data in Push-based Information Systems", *ACM SIGAPP Applied Computing Review*, 9(1):22–28, 2001.

In: Mobile Multimedia: Communication Engineering …   ISBN: 1-60021-207-7
Editors: I.K. Ibrahim and D. Taniar pp. 217-234   © 2006 Nova Science Publishers, Inc.

Chapter 11

LIMITED RESOURCE COMPUTATION

Daniel C. Doolan[*], *Sabin Tabirca*[**]

Department of Computer Science, University College Cork, College Road, Cork, Ireland

Laurence T. Yang[***]

Department of Computer Science, St. Francis Xavier University, Antigonish,
NS B2G 2W5, Canada

Abstract

Many modern day mobile devices such as phones are Java enabled allowing for Java 2 Micro Edition applications to be developed for them. The overwhelming majority of Java applications found on mobile devices are games, but these devices are capable of so much more. Many phones have processing speeds of 220Mhz such as the Nokia 6630 and 6680. So for such a small device it has significant computing capabilities. This chapter discusses the use of mobile devices to generate complex images that typically require significant processing to produce. It shows that current day devices are very much capable of carrying out significant computation. Mobile devices may also be used for other scientific problems, a classical example is Matrix multiplication, an $O(n^3)$ operation. We have all seen the Tricorders of Star Trek that are capable of being networked and used for all matter of tasks. Perhaps the Star Trek vision of mobile computing is already here. Together we shall explore some of these possibilities.

Introduction

Mobile computing is now ubiquitous, devices such as mobile phones and PDA's have infiltrated all of our lives. Just a year or two ago ordering a meal at a restaurant would require the waiter to write down the order using a pen and paper. This however is changing, the pen and paper is being replaced with devices such as a PDA and a stylus! Seeing a person talking to themselves on the street is now a regular occurance, as they are most likely making a phone

---

[*] E-mail address: d.doolan@cs.ucc.ie
[**] E-mail address: tabirca@cs.ucc.ie
[***] E-mail address: lyang@stfx.ca

call using their Bluetooth headset. Only a decade or so ago it would have been something that one would only see in a James Bond film. Children are now kept quiet for hours on end sending text messages to one another and playing games on these small pocket size devices. Are these devices capable of only fun and games, or do they have the potential to be essential scientific tools of the future?

Mobile Devices Are Everywhere!

The demand for mobile computing is ever increasing especially in relation to phones. No longer are phones capable of just text messaging, and voice communication. An entire plethora of capabilities now come as standard with a new modern day phone. These include video conferencing facilities, digital cameras, modem facilities to connect a Personal Computer to the Internet, the list is almost endless. Perhaps we have become a society infatuated with gadgets. In 2001 the usage of mobile phones in Ireland [13] stood at 67%, by the final quarter of 2005 it had reached 100% [14]. This is not just a single instance but the norm across many countries around the world. Several countries around Europe now have a 100% uptake in mobile phones. The highest currently stands at 156% (Luxembourg). Before long all of Western Europe will exceed 100% usage (perhaps as early as 2007) [19]. By the middle of the next decade (less than 10 years from now) it is expected that half the world's population will be mobile phone users [15].

Limited Resources

All computing devices have limited resources, but this is more prevalent for mobile devices. Even though devices such as mobile phones have very limited resources, they are not unlike the Desktop Personal Computer of ten years or so ago. They are in an ever evolving cycle just as Personal Computers and will only improve over time. Current day phones have several limiting factors to their usefulness: Screen Size, Memory, and Processing Power. Screen sizes vary in resolution, such as: 128 x 128, 128 x 160, 176 x 208, some of the high end communicator type phones have screen sizes as high as 640 x 200. In general the majority of Smartphones have a resolution close to 176 x 208 pixels. Available memory is another essential aspect for executing any program. The Nokia N71 [9] has 10MB of shared memory for storage, as do the Nokia 6630 and 6680. The amount of available memory is however increasing for example the Nokia N90 has 30MB of shared memory. The processing speeds are currently staying around the 200 MHz mark. The Nokia 6630 was their first 3G enabled phone, and yet more modern phones still use the same processor.

Java for Limited Devices

With such a large number of phones supporting a Java Virtual Machine it makes sense to develop for same to avail of this potentially huge market. Many of the lower end phones such as the Nokia Series 40 1st Edition provide support for Connected Limited Device Connectivity (CLDC) 1.0, and Mobile Information Device Profile (MIDP) 1.0. This restricts the mathematical computation to integers. To carry out the generation of the majority of fractal images requires the use of floating point (double) values. One can get around the problem of lower end phones that support only integer arithmetic by using the MathFP class that is widely available on the Internet. This class allows for floating point calculations using integer values, and so with this class it is possible to generate such fractal images as the Mandelbrot and Julia Sets. This can also prove of benefit as integer arithmetic is significantly faster to calculate that floating point, yielding faster computation of the image. To keep the operation involved in generating fractal images on a mobile device as straight forward as possible we will focus on using CLDC 1.1 and MIDP 2.0 which supports the use of double values. The majority of current phones are designed to use CLDC 1.1 and MIDP 2.0.

The Java libraries that are supported vary across devices. These are defined by the Java Community Process and are known as Java Specification Requests (JSR's). Examples of these are JSR 30 (CLDC) JSR 37 (MIDP). Many Smartphones have support for specifications such as: JSR 82 (Java API's for Bluetooth), JSR 120 (Wireless Messaging API), JSR 135 (Mobile Media API) and JSR 184 (Mobile 3D Graphics API for J2ME). Sony-Ericsson have a very clear structure to the JSR's that a particular mobile phone implements [17]. They divide the capabilities of their phones into six differing platforms, one through to six. Phones that implement the first platform have support for CLDC 1.0, MIDP 1.0 and JSR 135, while higher level platform have support for extra JSR's. Every increment of the platform strategy inherits all the features of the previous platform and adds a few more JSR's to the implementation. The highest level, platform six has support for CLDC 1.1, MIDP 2.0, JSR-185, JSR-120, JSR-135, Nokia UI API 1.1, JSR-172, JSR-205, JSR-184, Mascot Capsule v3 and JSR-75.

Phones as Capable Computers

We now move on to give some examples that demonstrate that mobile phones are capable of carrying out processor intensive tasks. The examples to follow are from the realm of both computer graphics and scientific computing. The primary focus will be on the generation of fractal images. This has be a favoured subject of mathematicians since the end of the 19th century. Up until the mid 1970's such images were typically referred to as "mathematical monsters". This all changed however with Benoit Mandelbrot who coined the term "fractal" from the Latin fractus or "broken". Mandelbrot infused the interest in fractals with his discovery of the Mandelbrot Set. The unique property of the Mandelbrot Set is its ability to index all the possible Julia Sets (Figure 1). The discovery of the Julia Set was a great achievement in the early part of the 20th century, but was almost forgotten until Mandelbrot's discovery. The generation of such images typically requires significant time to process even for desktop personal computers. We shall now explore the generation of several differing fractal images such as the Mandelbrot Set, Plasma Fractals and Prime Number Fractals. Finally we will show that phones are capable useful computation by the exploration of matrix multiplication an $O(n^3)$ problem.

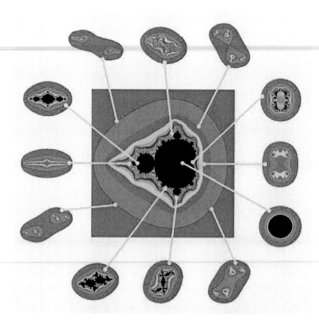

Figure 1: The Mandelbrot Set Indexing various Julia Sets

There are several different means by which a fractal image can be generated for display on a mobile device. It may be generated on a powerful server and sent to the phone for display over a network (using a http connection for example). Another approach is to use several mobile devices to carryout the processing, returning all sections of the image to a central device for display to the user. Such examples would typically use a Bluetooth network in the form of a point to multipoint Piconet to achieve this [5]. The final option is to generate and display the image on the phone itself [6]. This is the first method that will be examined. With such an interest in mobile phones by the majority of people, it is essential that these programs be developed in a user friendly manner. This would maximize usability and allow users to learn about this exciting area of geometry.

Generating the Mandelbrot Set

Fractal images are well known for their primary characteristics: self-similarity and infinite detail. The following figure (Figure 2) shows a typical example of the fractal zoom of the Mandelbrot Set. These images were generated on a desktop system using a well known colour map called "Volcano" that provides a rich and interesting colour gradient to the fractal image. The top right hand image is familiar to most people, the other images show the fractal zoom property, focusing in on the areas of greatest detail around the periphery of the image.

The Julia and Mandelbrot Sets are very similar in terms of both equations and generation functions. They use the polynomial function $f : C \rightarrow C, f(z) = z^u + c^v$ to generate a sequence of points $\{x_n : n \geq 0\}$ in the complex plane by $x_{n+1} = f(x_n), \forall n \geq 0$. The sequences $\{x_n : n \geq 0\}$ have two attractors 0 and infinity. The Julia and Mandelbrot sets retain only those initial points that generate sequences attracted by 0 as Equations (1 and 2) show

$$J_c = \{x_0 \in C : x_{n+1} = f(x_n), n \geq 0 \text{ are attracted by } 0\} \tag{1}$$

$$M = \{c \in C : x_0 = 0, x_{n+1} = f(x_n), n \geq 0 \text{ are attracted by } 0\}. \tag{2}$$

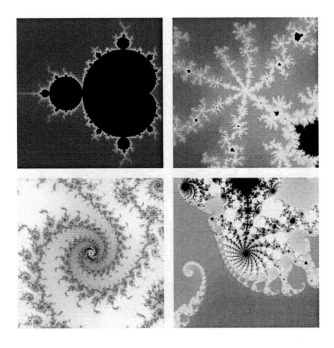

Figure 2: Mandelbrot (Z^2 + C), showing fractal zoom.

The generation of these fractal images is concentrated on a particular region of the complex plane between $[x_{min}, y_{min}] \times [x_{max}, y_{max}]$ and usually generate only the first n_{iter} points of the sequence x. If a point outside of a certain bound R e.g. $|x_n| \geq R$ then the sequence is not attracted by 0. To generate these fractals we need to calculate the first n_{iter} points for each point of the region and see whether the trajectory if finite or not. If all the trajectory points are under the threshold R we can then say that the initial point x_0 is in the set. The time complexity of the image generation process varies according to some factors, chiefly the image size, and the number of iterations for each point of the image.

The application carries out the generation of the fractal image in a Thread so event handling and such is not impeded. The fractal image is generated by modifying the values of a single dimensional array. Once the fractal generation procedure (Figure 3) has completed, the array is passed to the createRGBImage(…) method and an Image object is returned, which is then displayed on screen within a Canvas object.

```
procedure fractal
for each point (x,y) in [xmin, ymin]*[ymin,ymax] do
        construct the complex numbers c=x+j*y and z=0=j*0;
     for i=0 to niter do
         calculate z=f(z);
         if |z| > R then break;
     end for
draw (x,y) with the colour c[i%nrc];
end for
end procedure;
```

Figure 3: General Algorithm for Mandelbrot Set Generation

An Easy to Use Application

To allow for user control over the type of Mandelbrot image to be generated it proved necessary to have a simple to use interface (Figure 4) to modify the appropriate settings. The interface allows for the varying of the update rate, image size, the number of iterations, radius, powers for both Z and C and finally the equation type to use for the generation process. Another interesting option is "Invert Image" which inverts the C value of the Complex number and results in a very different looking image. This single option doubles the possible types of image that may be produced.

The formula type option allows the user to choose between three differing Mandel-like formulas: $Z_{n+1} = Z_n^u + C^v$, $Z_{n+1} = Z_n^u + C^v + Z$, $Z_{n+1} = Z_n^u - C^v$. A far greater variety of formulas (for example Table 1) could be produced if the Math class was more advanced. However had the application been implemented using the MathFP class then many other formula types could be possible.

Table 1: Some other possible Fractal Formulas

$Z_{n+1} = \cos(Z^u) + C^v$	$Z_{n+1} = \cos(Z^u) - C^v$
$Z_{n+1} = pow(C^v, Z^u) + C^v$	$Z_{n+1} = \cos(C^v, \exp(Z^u))$
$Z_{n+1} = \cos(\cos(Z^u + C^v))$	$Z_{n+1} = \cos(\cos(Z^u \times C^v))$
$Z_{n+1} = pow(\cos(C^v), Z^u)$	$Z_{n+1} = C^v + \cos(\exp(Z^u))$

The "Update Rate" option allows for the screen to be repainted after a certain number of columns of the image have been generated. This allows the user to see the fractal image being generated. Setting the "Update Rate" to the same value as the image size will result in the application executing the generation process as fast as possible. Only when the image has been completely generated will it be displayed on screen. This was the option used for testing of the application.

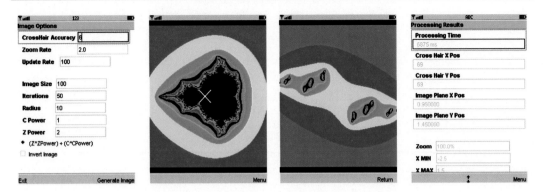

Figure 4: Options, Mandelbrot Set, Julia Set and Results, Screens

The Canvas on which the Mandelbrot Image is displayed also has a crosshair (Figure 4) that may be moved according to user input. The user can then select an option from the menu to view the corresponding Julia set at the point indicated by the crosshair. This will result in another Thread being initialised to generate and display the corresponding Julia Set. The user may also zoom in on any area of the Mandelbrot set, focused about the area indicated by the crosshair. The rate of the zoom in feature is controlled from the initial options screen "Zoom Rate" parameter.

Processing Times

Testing the Mandelbrot generation application on both the Nokia 6630 and 6680 gave similar results. In the case of generating the image for 50 iterations the Nokia 6630 carried out the processing slightly faster in all test that the 6680. The reverse is the case with regard to the generation of an image for 500 iterations. The tests also showed that the 6630 was capable of generating an image 600 x 600 pixels while the 6680 was unable to (Tables 2, 3, Figures 5 and 6). The 6630 and 6680 are both quite high end phones in terms of processing power. Running the program on a far less sophisticated phone such as the 3320 produced limited results. The 3320 was unable to generate an image of 200 x 200 pixels, however it was capable of generating an image 150 x 150 pixels. To do this required 119,360ms at 50 iterations and 696,075ms at 500 iterations, clearly far too long to be of any practical use.

Table 2: Mandelbrot Image Generation Processing Times at 50 Iterations

Device	500×500	400×400	300×300	200×200	100×100
Nokia 6630	53,359	33,812	19,484	9,812	3,000
Nokia 6680	60,859	36,234	21,859	12,516	3,141

Table 3: Mandelbrot Image Generation Processing Times at 500 Iterations

Device	500×500	400×400	300×300	200×200	100×100
Nokia 6630	301,344	196,125	111,484	52,344	13,281
Nokia 6680	290,062	187,797	103,219	48,000	12,359

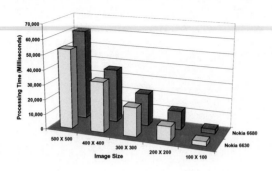

Figure 5: Mandelbrot Image Generation
Processing Times at 50 Iterations

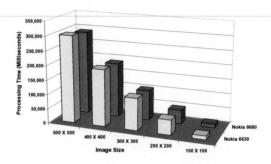

Figure 6: Mandelbrot Image Generation
Processing Times at 500 Iterations

Generating the Prime Number Fractal

Prime numbers have remained a topic of interest for well over two and a half thousand years. References to this topic of mathematics have been traced back all the way to its birth place in ancient Greece and the time of Euclid. To this day a strong interest in the topic still remains. One area clearly evident on the Internet is the search for ever larger prime numbers. The past year or so has seen ever larger prime numbers discovered (Table 4) principally in part to the GIMPS (Great Internet Mersenne Prime Search) project [7]. The hunt is on for a prime number that contains 10 million digits or more. At the current rate of progress researchers believe this should be found within the next year or so.

Table 4: Most recently discovered prime numbers

Discovered	Rank	The Prime	# Digits	Discoverer
15th December 2005	43rd	$2^{30,402,457}-1$	9,152,052	Dr. Curtis Cooper and Dr. Steven Boone
18th February 2005	42nd	$2^{25,964,951}-1$	7,816,230	Dr. Martin Nowak
15th May 2004	41st	$2^{24,036,583}-1$	7,235,733	Josh Findley

The development of methods to visualise primes has been a subject of interest for quite some time. By far the most well know example of this was discovered by Stanislaw Ulam (the Ulam Spiral) in 1963 [18]. The Ulam Spiral was created by placing a zero in the middle of a grid and incrementing the number at each new cell around the central area in a spiral fashion. Highlighting the prime numbers leads to the visualisation of the Ulam Spiral. The interesting aspect about this image is that the primes seem to produce straight lines. The prime number fractal (Figure 8) is another method of visualising primes, it being first constructed by Adrain Leatherland of Monash University Australia [8]. The generation algorithm (Figure 7) iterates through a sequence of primes. A direction is assigned by carrying out modular division. The direction is used to move to the next cell where its colour is incremented.

```
procedure pnf_fractal
for (i=1; i<sieveSize; i++){
          if (isPrime(i)){
                dir = prime %5;
                if(dir ==1)  x --
                if(dir ==2)  x ++
                if(dir ==3)  y --
                if(dir ==4)  y ++
                incrementColour(x,y)
          }
     }
end procedure;
```

Figure 7: Prime number fractal generation algorithm

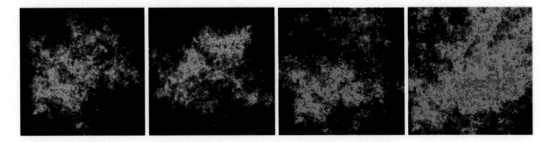

Figure 8: Examples of the Prime Number Fractal

The development of a mobile application to achieve the generation of a prime number fractal requires an initial Form object to allow for the assigning of user designated properties. The data fields required include: Sieve Size, Image Size and starting coordinates for the x and y position (Figure 9). With appropriate parameters selected the fractal generation process may be initialized resulting in the generation of a prime number fractal image that reflects the settings selected by the user (Figure 9).

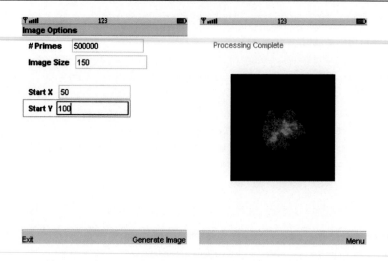

Figure 9: User settings screen and generated image screen

The dominant feature of prime number fractals is the central area of brightness. This feature is a result of the pixels within the central area being visited (and incremented) far more often than pixels around the periphery of the image. The trajectory that a move will take occurs randomly but the total number of moves in any direction stay approximately the same. Typically the deviation in the number of moves (even when several million primes are generated) is often only a few hundred of a difference.

Theorem 1. The numbers of *up, down, left*, and *right,* moves in the PNF algorithm are asymptotically equal.

Proof. Dirichlet's theorem assures if a and b are relatively prime then there are an infinity of primes in the set $\{a \cdot k + b, k > 0\}$. This means that the random walk has an infinity of *up, down, left,* and *right,* moves. If $\pi_{a,b}(x)$ denotes the number of primes of the form $a \cdot k + b$ less than x then we know from a very recent result of Weisstein [20] that

$$\lim_{x \to \infty} \frac{\pi_{a,b}(x)}{li(x)} = \frac{1}{\varphi(a)} \tag{1}$$

where $li(x)$ is the logarithmic integral function and $\varphi(a)$ the Euler totient function. The particular case $a = 5$ gives

$$\lim_{x \to \infty} \frac{\pi_{5,k}(x)}{li(x)} = \frac{1}{\varphi(5)} = \frac{1}{4}, \forall k \in \{1,2,3,4\}, \tag{2}$$

which means that

$$\pi_{5,1}(x) \approx \pi_{5,2}(x) \approx \pi_{5,3}(x) \approx \pi_{5,4}(x) \approx \frac{li(x)}{4} \tag{3}$$

Equation (3) shows that the 2D PNF algorithm has asymptotically the same number of *up, down, left,* and *right* moves (Table 5).

Table 5: Distribution of Moves for the PNF Image.

#Primes	20,000	40,000	60,000	80,000	100,000	10,000,000
Left	4,978	10,003	14,971	19,985	24,967	2,499,755
Right	5,023	10,013	15,020	20,007	25,016	2,500,284
Up	5,011	9,997	15,018	20,031	25,007	2,500,209
Down	4,987	9,986	14,990	19,976	25,009	2,499,751

The generation of the primes is carried out using the sieve of Eratosthenes's. This generates all the prime numbers $p_0 = 2, p_1 = 3, ..., p_{m-1}$ less than n. The Prime Number Theorem gives that $m = O(\frac{n}{\log n})$. The Eratosthenes' sieve provides an $O(n \cdot \log \log n)$ computation and that is recognized to be the most efficient algorithm to generate these primes when the number of primes is small. Since this is the most expensive computation of the image it results that the overall complexity is $O(n \cdot \log \log n)$ [3].

The generation of the image is a simple process where by once a prime has been found it is mapped to a new cell location and the colour of that cell is incremented. The mapping of primes to directional movement is carried out by $p \bmod 5 \in \{1,2,3,4\}$, yielding four possible moves (Table 6).

Table 6: Mapping of resultant modular values to movement

Result	Direction	Mapping
1	Left	$p \bmod 5 = 1 \Rightarrow (x,y) \Rightarrow (x\text{-}step,y)$
2	Right	$p \bmod 5 = 2 \Rightarrow (x,y) \Rightarrow (x\text{+}step,y)$
3	Up	$p \bmod 5 = 3 \Rightarrow (x,y) \Rightarrow (x,y\text{-}step)$
4	Down	$p \bmod 5 = 4 \Rightarrow (x,y) \Rightarrow (x,y\text{+}step)$

Processing Results

The generation of prime number fractals on mobile phones seems to be quite limited (Figure 10). The generation of the prime numbers is the most expensive part of the image generation process. Even the generation of just 100,000 primes requires significant time well over 500,000ms for both the 6630 and 6680 phones. Based on this the generation of an image based on a few million primes in reasonable time is currently beyond the capabilities of today's phones. The sieve size required even for the generation of only 100,000 primes was far greater than the number of primes actually produced (Table 7).

Table 7: Sieve Sizes required for testing

# Primes	20,000	40,000	60,000	80,000	100,000
Sieve Size	224,740	479,930	746,775	1,020,380	1,299,720

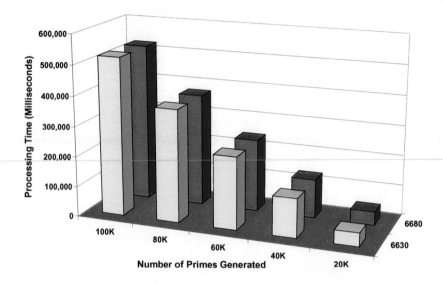

Figure 10: Prime Number Fractal Generation Times

Generating Plasma Fractals

This variety of fractals is often referred to as "Fractal Clouds" due to the cloud like nature of the resulting image (Figure 11). The procedure for the generation of this type of image differs greatly to that of the Mandelbrot Set. For the plasma fractal the image is created using a recursive algorithm known as random midpoint displacement. Using this algorithm in the two dimensional domain for displacing colour values will result in plasma fractals. However applying the very same algorithm to height values in a three dimensional domain will yield randomly generated terrain (Figure 12). This method of terrain generation was used by the film industry for the creation of an alien moon in the film "Start Trek II the Wrath of Kahn".

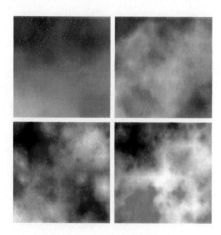

Figure 11: Plasma Clouds

Figure 12: Fractal Terrain

The recursive method (Figure 14) to generate the fractal firstly requires the colours for each corner of the image to be assigned. Running the recursive algorithm will examine the colour values at each corner average them, and then displace the averaged value. The displacement amount is derived from the width and height of the quadrant. This averaged value is the colour that is assigned to the central point of the image. Once this operation has completed, the method is recursively called again four times. On each new call a subsection of the image is processed. This means that the image is continually subdivided into four quadrants until the pixel level is reach. Once the pixel level is reached the subdivision process can no longer be carried out and the fractal generation process is terminated.

Similar to the previous example (Prime Number Fractal Generation) the user interface for the creation of plasma images is very straight forward. Essentially only two parameters are necessary to generate this type of image: the grain size and the image size. The resultant generated image is displayed within a Canvas object (Figure 13).

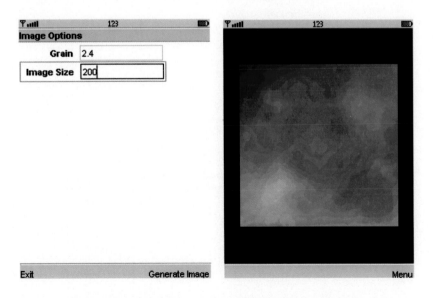

Figure 13: User settings screen and generated image screen

```
divide(x,y,w,h,tLC,tRC,bRC,bLC)
begin
    float nW = w /2, nH = h /2;
    if(w > 1 OR h > 1)
        int displace = displace(nW,nH);
        Color top =      avgColors() + displace;
        Color right =    avgColors() + displace;
        Color bottom = avgColors() + displace;
        Color left =     avgColors() + displace;
        Color centre = avgColors() + displace;
        divide(x,y,nW,nH,tLC,top,centre,left);
        divide(x+nW,y,w,h,top,tRC,right,centre);
        divide(x+nW,y+nH,w,h,centre,right,bRC,bottom);
        divide(x,y+nH,w,h,left,centre,bottom,bLC);
    else
        drawPixel(x,y)
end
```

Figure 14: Random Midpoint Displacement

Processing Results

The plasma fractal is by far the fastest to generate of the three fractals presented in this chapter. The generation of a 400 x 400 pixel image required only 6,328ms on the 6630, and 23,453ms for a 600 x 600 pixel image (Figure 15).

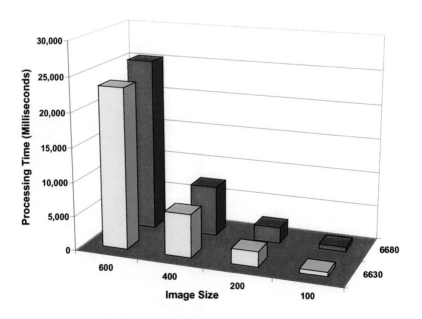

Figure 15: Plasma Fractal Generation Times

Matrix Multiplication

This last section shows that mobile devices are capable of performing useful number crunching. This can be seen in the use of a mobile phone for matrix multiplication an $O(n^3)$ operation. The product of two matrices A and B may be defined as: $c_{ik} = \sum_{j=1}^{m} a_{ij} b_{jk}$, where j is summed for over for all possible values of i and k. The dimensions of the matrices must satisfy $(n \times m)(m \times p) = (n \times p)$. More efficient algorithms do exist however for multiplying two $n \times n$ matrices all cells n^2 must be processed, hence it cannot run faster than $O(n^2)$. Strassen's algorithm [16] gives a time complexity of $O(n^{2.807})$. The best known algorithm today (Coppersmith-Winograd algorithm [4]) has a complexity of $O(n^{2.376})$.

A simple to use user interface (Figure 16) was developed using the standard GUI components provided by J2ME. The matrices A and B could be assigned a predetermined value or a randomly generated value, with little to no impact on the final results. The other options allow for the varying of the matrix size (square matrix) and also the data type (integers, floats, doubles). Once the process of matrix multiplication is complete the processing times for same are presented to the user. To verify correct computation the user may query any cell of the computed matrix by giving the cell location. The method (Figure 17) to carryout the multiplication process is quite straightforward essentially containing just three for loops, to loop through all the required cells.

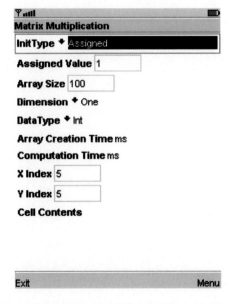

Figure 16: Matrix Multiplication User Interface

```
private void product(int n, int m, int p, int[][] a, int[][] b, int[][] c){
    int i,j,k;
    for(i=0;i<n;i++)for(j=0;j<p;j++){
        c[i][j] = 0;
        for(k=0;k<m;k++){
            c[i][j] += a[i][k] * b[k][j];
        }
    }
}
```

Figure 17: Example of generating the product of two matrices $O(n^3)$

Processing Times

Again the application was tested on both the 6630 and 6680 phones. The upper limit of matrix size that could be computed for integers was 700×700, the memory requirements being well over 5.5MB with a processing time of 364,813ms. In the case of double value the upper limit was a matrix of 490×490 and required 308,171ms to process. Graphical representations of the processing times for several sized matrices may be seen in Figures 18 and 19. In general for matrices of 100 to 300 in size the processing times of doubles was approximately 4:1 that of integers, reducing to about 3:1 for a matrix of size 400.

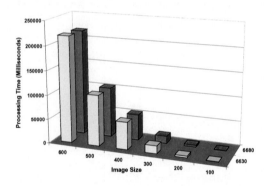

Figure 18: Processing Times for Matrix Multiplication of Integers

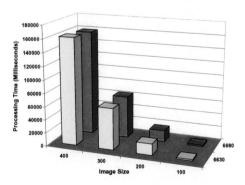

Figure 19: Processing Times for Matrix Multiplication of Doubles

Conclusion

Several examples of computationally expensive tasks have been explored. It has been shown that mobile phones are quite capable of performing such tasks in a reasonable amount of time. The tests showed that the test phones were capable of performing matrix computation on matrices of size 700^2. The generation of fractal images such as the Mandelbrot Set, Julia Set, and Plasma fractal were all quite successful. Fractal images of up to 600 x 600 pixels were generated. The least successful tests dealt with the generation of the prime number fractal. The time required to generate the primes being the limiting factor. The examples were tested on phones that have a 220Mhz processor, and 10MB of shared memory.

Several phones today already have far greater memory capacities such as the Nokia N70. The processing speeds will also increase over the coming years allowing for far more complex tasks to be performed. This is clearly evident with the announcement of a 1Ghz Cortex-A8 processor by ARM (4th October 2005) [1][2][12].

With such processing power becoming available the phones of a few years from now will have capabilities approaching the desktop machines we have today. It would seem that the Star Trek vision (circa 23rd and 24th centuries) of hand held portable computers (Tricorders) is far closer than we may thing. Mobile phones are in a process of evolutionary change with additional features being added all the time. Only a few years ago the latest technology enabled text messaging and voice communication. What technologies will be available even ten the years from now?

Currently one could argue that a mobile phone is more a digital camera than a phone. The Nokia N92 [11] is capable of receiving live digital TV broadcasts, has a built in FM radio, capabilities for playing audio files, two mega pixel camera and video recorder. No longer is a phone just a phone, but it has been transformed into a multimedia centre capable of handling just about all the various media formats that a standard computer today is capable of.

References

[1] ARM, ARM Cortex-A8, http://www.arm.com/products/CPUs/ARM_Cortex-A8.html, 2005.
[2] ARM, ARM Introduces Industry's Fastest Processor for Low-Power Mobile and Consumer Applications, http://www.arm.com/news/10548.html, Oct, 2005.
[3] E. Bach, J. Shallit, Algorithmic Number Theory, Vol 1: Efficient Algorithms, MIT Press, 1996.
[4] D. Coppersmith, S. Winograd, Matrix multiplication via arithmetic progressions, *Journal of Symbolic Computation*, **9**:251-280, 1990.
[5] D. Doolan, S. Tabirca, Distributed Fractal Generation Across a Piconet, Proceedings of Sidrad05 – Mobile Computer Graphics Conference, Lund, Sweden, November 2005, pp 63-68.

[6]   D. Doolan, S Tabirca, Interactive Teaching Tool to Visualise Fractals on Mobile Devices, Proceedings of the 6th Irish Eurographics Workshop on Computer Graphics, Dublin, Juen 2005, pp 7-12.

[7]   GIMPS, Great Internet Mersenne Prime Serach, http://mersenne.org.

[8]   A Leatherland "Pulchritudinous Primes: Visualising the Distributon of Prime Numbers", http://yoyo.cc.monash.edu.au/~bunyip/primes.

[9]   Nokia, Nokia N71 Technical Specs, http://www.forum.nokia.com/main/0,,018-2815,00.html?model=N71, November, 2005.

[10] Nokia, Nokia N90 Technical Specs, http://www.forum.nokia.com/main/0,,018-2579,00.html?model=N90, April, 2005.

[11] Nokia, Nokia N92 Technical Specs, http://www.forum.nokia.com/main/0,,018-2776,00.html?model=N92, November, 2005

[12] F. Pilato, ARM Reveals 1Ghz Mobile Phones Processors, http://www.www.mobilemag.com/content/100/102/C4788/, Oct, 2005.

[13] RTE, Internet and Mobile Penetration Still Rising, http://www.rte.ie/business/2001/0308/odtr.html, 2001.

[14] RTE, Mobile Penetration Now Stands at 100%, http://www.rte.ie/news/2005/1220/mobilephones.html,2005

[15] RTE, Half the world to have mobile phones by 2015, http://www.rte.ie/business/2004/0225/phones.html, 2004.

[16] V. Strassen, Gaussian Elimination is not Optimal, Numer. Math. 13, 254-356, 1969.

[17] Sony-Ericsson, Java Platform Versions, http://www.sonyericsson.com/developer images/javaplatformversionsandscreensizesjune2005.xls, June, 2005

[18] H. Systems, Prime Number Spiral, http://hermetic.ch/pns/pns.htm.

[19] A. Taylor, Mobile penetration to hit 100% in Europe, http://www.enn.ie/news.html?code=9604990, 2005

[20] E. Weisstein , Arbitrarily Long Progressions of Primes, *MathWorld* headline news, http://mathworld.wolfram.com/news/2004-04-12/ primeprogressions/, April 12, 2004.

In: Mobile Multimedia: Communication Engineering …   ISBN: 1-60021-207-7
Editors: I.K. Ibrahim and D. Taniar pp. 235-257   © 2006 Nova Science Publishers, Inc.

Chapter 12

RADIO RESOURCE MANAGEMENT FOR QOS PROVISIONING IN COMMUNICATIONS NETWORKS BEYOND 3G

Ángela Hernández-Solana[*] *and Antonio Valdovinos-Bardají*[**]
Electronics Engineering and Communications Dpt.
Institute of Engineering in Aragón, University of Zaragoza
C/ María de Luna 3, 50018, Zaragoza, SPAIN

Abstract

This chapter presents an overview of Radio Resource Management (RRM) schemes for supporting of multimedia applications with various Quality of Service (QoS) requirements in IP CDMA-based systems. The revision emphasizes particularly the issues and the approaches related to the design of the call admission control (CAC) for the fourth-generation (4G) wireless networks. In contrast to traditional circuit switched cellular wireless networks, 4G will be based on packet switching at the wireless interface and will inter-work with IP based wired networks. The study concentrates on the proposals of enhanced versions of the air interface based on the WCDMA (Wideband Code Division Multiple Access) access technique, defined in the UMTS (Universal Mobile Telecommunication System) system, although the study can be generalized to any cellular system based on the CDMA access technique. If in all wireless mobile networks radio access remains the most challenging component to provide QoS bearer services, QoS provision becomes even a more challenging task when the radio interface being in transition from circuit to packet based networks and particularly for the WCDMA uplink interface. Call admission control schemes will be a key component and a difficult task in these systems. Packet transmission combined with the WCDMA access technique provides high flexibility in resource allocation in order to integrate multimedia services with different QoS requirements, but makes CAC a more complex problem than in traditional cellular networks due to the traffic variability (with different QoS constraints) and to the fact that capacity is not fixed and varies with stochastic changes in the interference level.

[*] E-mail address: anhersol@unizar.es
[**] E-mail address: toni@unizar.es

Introduction

As mobile communications and Internet converge, packet based multimedia services including IP (Internet Protocol) based services, such as IP telephony, are expected to be a dominant traffic component on the Third/Fourth Generation (3G/4G) wireless networks. Packet based evolutions of the currently developed 3G cellular access systems (such as the UMTS system), are believed to be significant segments of future global wireless IP networks, that shall include IEEE 802.16/802.20 metropolitan area networks, IEEE 802.11 wireless local area networks (WLANs), Bluetooth/ultra-wideband (UWB) personal area networks (PANs) and 3G/4G mobile telephone systems in addition to wired networks. In this context, most of the current research efforts are conducted to extend the capacity and capabilities of the 3G cellular wireless networks, developing the potentialities of packet switching schemes at the wireless interface in an efficient way towards wireless IP networks capable of providing Quality of Service (QoS) bearer services which could support multimedia applications with real time constraints and variable bit rates in addition to the current best-effort services.

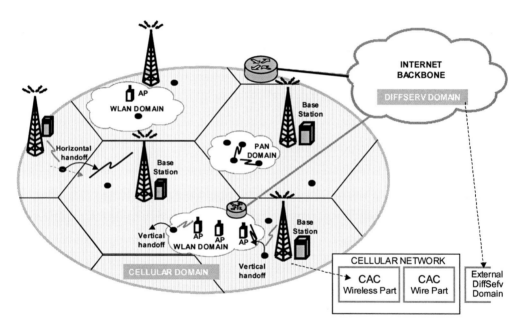

All IP heterogeneous wireless networks.

In any case, although in the proposed architectures the transmission bottleneck is the wireless link, provision of end to end QoS in an IP heterogeneous network makes it necessary not only to manage the QoS within each domain, but also to consider end-to-end QoS provision as a global process with close interrelation between the individual domains.

In the wire-line IP context, the significant amount of research efforts carried out in the last years to that issue culminated in two completely different networking frameworks in order to provide end-to end QoS: Integrated Services (IntServ) [1] and Differentiated Services (DiffServ) architectures [2]. The IntServ is a flow-based QoS architecture that provides a certain service guarantee to each individual traffic flow. Resources for traffic are explicitly identified and reserved, so, one of the key requirements is how QoS routing identifies efficient paths that can satisfy the given QoS constraints taking into account the application requirements and the availability of network resources. In this case, explicit dynamic resource reservation, in conjunction with admission control, is used. On the other hand, in the DiffServ architectures, traffic is differentiated into a set of classes (premium, assured forwarding and best effort) and network nodes provide priority-based treatment of these classes. In this case, no explicit resource reservation or admission control is employed and, therefore, differentiate treatment is relative. In addition, DiffServ eliminates the need for processing per-node that IntServ requires, reducing the scaling problems of IntServ in large networks, and allows it to adopt a domain-based architecture where each domain can be independently chosen. Note that IntServ requires implementing a resource reservation protocol (i.e. RSVP) in all networks segments from the source to the destination in order to reserve resources before starting transmitting data. All these aspects make DiffServ the most promising architecture to the global network, given that it provides the flexibility to develop or modify techniques in each domain without a significant impact in other domains. A desirable approach is to consider at least two independent domains with different conditions and problems, one for the wired part and the other for the packet based evolution of the wireless part.

Restricting the scope of the problem to a specific wireless access, if we consider, for example, the establishment of a multimedia application over the Internet using the UMTS system as the access network for at least one of the end user terminals, we can define, from the point of view of the UMTS system, two different levels of QoS in order to satisfy the QoS end-to-end service, each providing the respective bearer service (BS) between two defined points: an External BS and a UMTS BS [3][4]. In this context, the IP BS manager uses standard IP mechanisms to control the external IP service. Really, it may include support of DiffServ and RSVP functions. Given that these mechanisms may be different from techniques used to maintain QoS within the UMTS domain, it is required the IP BS manager to provide the inter-working between the mechanisms and the parameters used within the UMTS BS and those used within the IP BS (translation function). Moreover, the interaction between the UMTS BS and the IP BS shall be limited at this translation function between attributes and requests. Obviously, this separation is performed in order to facilitate a separate evolution of both IP and UMTS traffic policy mechanisms.

The end-to-end Bearer Service is out of the scope of this chapter, which focuses on the UMTS BS (Core plus Radio Access Bearer Services) that provides the UMTS QoS and particularly in the Radio BS. QoS provisioning in the Radio Access remains to be one of the most challenging problems for 3G cellular-based wireless networks. The characteristics of the wireless link (limited bandwidth, unstable channel quality and interference) and the users mobility make the achievement of this QoS requirement more complicated than in wired networks. On the other hand, although packet transmission combined with CDMA (Code Division Multiple Access) access technique used in UMTS can potentially improve the system efficiency in front of traditional TDMA systems, a more complex and sophisticated radio resource management will be required, since that capacity is not fixed and varies with stochastic changes in the interference level. Interference handling by radio resource allocation schemes plays an important role to enhance the performance and particularly to improve the capacity in the uplink.

Radio BS covers all the aspects of the radio interface transport that is provided by the UTRAN (UMTS Terrestrial Radio Access Network) in the FDD (Frequency Division Duplex) mode. A collection of RRM algorithms guarantees the stability of the radio path and the QoS of the radio connection by efficient sharing and managing of the radio resources. Differentiated allocation of resources through *power control*, prioritized access to medium through *packet scheduling* and specially *admission control* and performance guarantees during *handoff* will be discussed in the rest of the chapter related with mobility characteristics and traffic patterns associated to the new services. Essentially, QoS can be provided by a combination of call admission control (CAC) and flow control. While CAC indirectly controls interference by limiting the number of users in the system, in a packet time scale, the medium access/flow control can balance the system interference and arranges transmissions on a frame-to-frame basis, specially for the WCDMA uplink interface. Note, however, that although the study takes network UMTS as a reference, the results and the exposed considerations can be generalized to any cellular system based on the CDMA (Code Division Multiple Access) access technique where transmission of multimedia services is performed in packetized form.

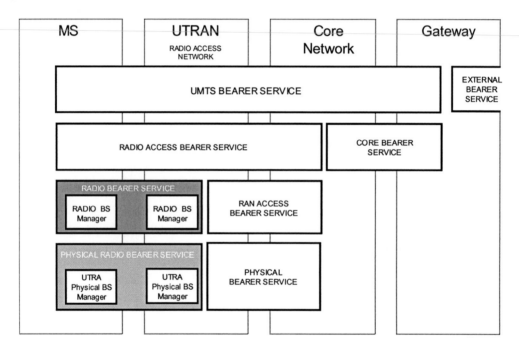

UMTS QoS Architecture.

Due to the nature of difficulties in dealing with wireless radio channels, the separation between Radio and Core Bearer Services is required. QoS mechanism definition must allow independent evolution of Radio and Core networks in order to eliminate or minimize the impact of the evolution of transport technologies in the wire line world over wireless access.

Considering the large set of possible applications supported by UMTS grouped in four main categories, where the main distinguishing factor is how delay sensitive the traffic is: Conversational (very delay sensitive), Streaming, Interactive and Background (the most delay insensitive) classes, within Radio Resource Management (RRM) strategies it is possible to consider combinations of IntServ and DiffServ approaches. DiffServ can be used, for example, in order to prioritize conversational or streaming traffic in front of background classes. However, in any case, this must be accompanied by a tight control of the traffic, admission control, in order to ensure that the carried traffic is within the capacity of the radio access network. A session based approach needs to be adopted for all packet mode communication, while sufficient detail about the applications traffic and service requirements must be known in order to determine the appropriate QoS attributes to enable service optimization. Radio Access Bearer assignment request is investigated by the RRM admission control algorithm, which checks whether the Radio Bearer (RB) can be established with the requested QoS.

In addition to the efficient design of QoS control mechanisms working at the physical (power control and interference management), link (prioritized access to medium through *packet scheduling)* and call levels, the problem is how to map the large set of parameters (power, spreading factors, Bit and Block Error Rate (BER/BLER), delay and delay jitter) and aspects (mobility, channel instability, variant resource availability due to the interference) which control or condition the characteristics of the Radio BS into a limited set of resource reservation parameters to be used by the admission control. A cross-layer design of CAC should be applied to capture both packet level and call level QoS performance measurements.

If Radio BS is allowed to be established with the QoS parameters, the UMTS BS is ready to carry data flows according to the end-to-end QoS requirements. In a general sense, we could think about an end-to-end bearer service that could or could not include DiffServ regions in the External domain, in the middle of a larger network supporting IntServ approaches (resource reservation and call admission control) at the wireless access interface. CAC at the DiffServ edge router should be able to negotiate an appropriate service level agreement (SLA) with the DiffServ domain, so the decision to finally accept or reject a call should be based on both the availability of wireless resources and the negotiated SLA. Since the wireless part has the scarcest resources in the system, the wired part must ensure that the traffic already transmitted across the wireless links can be maintained. Conversational and streaming call can be mapped into premium and assured forwarding service classes, respectively, while non-real time services can be mapped into the best effort service calls.

Challenges in Resource Management to QoS Provisioning over CDMA Radio Access Interface

If in all wireless mobile networks radio access remains to be the most challenging component to provide QoS, this provision becomes even more challenging task when the wireless mobile network evolves from circuit to packet based transmission and particularly when the air interface is based on CDMA. QoS must be controlled in three different levels: physical (BER and BLER rates), packet/link level (packet transmission delay and delay jitter, and packet dropping probabilities), and call level (call blocking and dropping probabilities). Ensuring the appropriate QoS requires an efficient coordination and a more complex and sophisticated design of the actual RRM mechanisms, particularly call admission control and flow/load control, in addition to the power control.

In interference limited systems, as CDMA, power control is one of the basic requirements. Any reduction in the interference converts directly into an increase in the capacity, and therefore, achievement of both transmission rates and delay requirements is closely related directly or indirectly to power allocation. The effects of mutual interference

(MAI) among active connections to the QoS provision is specially important in the uplink, since, as the number of active users increases the interference causes a rising degradation of connections that, if uncontrolled, prevents from guaranteeing the necessary QoS at the physical and link levels. Thus, good interference handling by radio resource allocation schemes plays an important role to enhance the performance and to increase the system capacity. In this way, call admission control indirectly controls interference by limiting the number of users in the system, performing an important role as a procedure for congestion control, particularly when connection oriented services are considered. At the same time, on a lower level, the flow control handles packets from different sources according to their QoS requirements (delay, rate). The correlation between the two levels is obvious, so the effects of a good coordination in the system capacity are clear. If too much traffic is allowed to enter the network by a lax CAC, no flow control/scheduler will be able to provide the requested QoS and, otherwise, the achievement of a high level of utilization also depends on the ability of the CAC algorithm to minimize the call blocking and call dropping probabilities. A CAC algorithm that unnecessarily denies access to users that could have been successfully admitted will underutilize the network resources.

In general, cross-layer optimization can lead to a significant performance improvement. CAC cross-layer design should be applied to capture both call and packet-level QoS performances. In fact, CAC can be seen as a control which really operates at two different levels: the first one characterized by the "packet level" constraints, such as packet loss, delay jitter or average delay and, the second one, which allows to share the system capacity among the various traffic types and/or to protect handoff calls from the new ones. Due to the seamless connection and global mobility requirements in a 4G system, in addition to traditional handoff calls between adjacent cells (horizontal handoff), calls from another particular network (for instance a WLAN network) must be able to transparently roam to the cellular network (vertical handoff), resulting, from the CAC point of view, in a subtype of handoff call. A new performance metric, vertical handoff call dropping probability, should be determined, despite the fact that "bandwidth" or parameters adaptation will be required between networks as explained in the introduction.

RRM is required in both uplink and downlink interfaces. Nevertheless, while the uplink is interference limited, the downlink is power limited. In the downlink, the total transmitted power in a Radio Frequency (RF) carrier is shared between the users receiving from the Base Station (BS), whereas in the uplink, there is a maximum tolerable interference level at the BS receiver that is shared between the transmitting mobile stations in the cell, each contributing to the interference. The differences between the uplink and the downlink require a separated analysis for both links. However, the description carried out in this chapter is restricted almost exclusively to the uplink, since uplink control could be considered a more challenging issue. On the other hand, although the discussed improvement proposals, described next, associated to the RRM mechanisms in the transmission protocol stack can be generalized to any CDMA system, the references to UMTS throughout the chapter are particularly important when we refer to the organization of communications, definition of logical and physical channels and its implications in the configuration of the access control as a method to the management of the transmissions.

Power and Rate Allocation

In a CDMA system, both power and transmission rates may be considered as controllable resources. It is widely known that the requirements associated with a mobile station (MS) i , located in a cell k, in terms of maximum bit (BER) or block (BLER) error rates, when this mobile is connected at a data rate, $r_{i,k}$, can be mapped into an equivalent required constraint of the ratio signal energy per bit (E_b) over total interference density (N_o) received at the Base Station (BS), denoted by γ_i in expression 1):

$$\left(\frac{E_b}{N_o}\right)_{i,k} = \frac{W}{r_{i,k}} \cdot \frac{P_{i,k} \cdot h_{i,k}}{I_{intra,k} + I_{inter,k} + \eta_o \cdot W} \geq \gamma_i \quad i = 1..N \qquad 1)$$

where $I_{intra,k}$ and $I_{inter,k}$ are the intra-cell and inter-cell interference powers received at the base station (BS), respectively; η_o is the thermal noise spectral density, $P_{i,k}$ the transmitted power associated to the MS i in cell k, $h_{i,k}$ the path loss between the MS i and the BS k and W the available bandwidth in the cell (chip rate). In summary, BER and BLER guarantees can be achieved by satisfying the signal to interference ratio (SIR) or $(E_b/N_o)_{req,i,k}$ requirements, and these are obtained by power control. Originally, in a context of homogenous traffic networks, most of existing works assume that power control was oriented to overcome the "near-fair" effect. Power regulation takes place in such a way that the power received at the BS was the same for each connection, guaranteeing therefore the same quality. However, power control imperfections may occur occasionally and cause maladjustments in the received power. Nevertheless, in the context of heterogeneous services, a more effective and suitable power control meets differentiated $(E_b/N_o)_{req,i,k}$ requirements to each of the individual users in order to guarantee their specific BER and BLER constraints associated to a given rate transmission requirement.

In the uplink, power control is implemented as a combination of open loop and closed loop. Through frequent update commands, the BS attempts to control the received power from each mobile terminal. Difficulty on power allocation and adjustment requires that RRM incorporate more accurate estimations of power control imperfections.

In fact, not only power control is needed in order to mitigate variant wireless channel conditions and to maintain the required $(E_b/N_o)_{req,i,k}$ associated to specific BER requirements, but also an optimum power assignment criterion is demanded in order to limit both intra-cell and inter-cell interference levels. There exist a variety of power assignment criteria, although in general it is desirable to minimize the sum of powers. In fact, the purpose of the minimum power control criterion (MPCC) proposed and described in [5] is to meet the BER for all the simultaneously transmitting users, assigning an optimum level to the power transmitted by each user in such a way that interference caused to other cells is minimized and so, throughput is maximized. Besides, the interference level at the BS is always maintained under a threshold, while output powers of mobile stations are constrained.

Joint power and rate assignment is one of the main subjects of RRM. The flow control and packet scheduling, commonly associated to medium access control (MAC) functionalities at the Base Station (BS), in coordination with an optimum power criterion, should be responsible for arranging the transmission of packets within their specified rate requirements and delay tolerances taking into consideration the channel conditions of the individual users. Different data rates are supported on the traffic channels through the use of different spreading factors or through multicode transmission, while, in general, the assumption of a centralized demand assignment protocol at the BS should lead to a more accurate QoS control. Several scheduling strategies, based on static and dynamic priorities, could be used to exploit service burstiness and delay tolerances, accommodating more users and maximizing resource utilization. In any case, the optimum power allocation, frame to frame, is only possible if:

$$\sum_{i \in Cellk} C_i = \sum_{i \in Cellk} \frac{1}{\left(\dfrac{W}{r_{ik}\gamma_i}+1\right)} \leq \frac{1-\eta(t)}{(1+f(t))} = C_{max}(t) \qquad 2)$$

where $f(t)$ represents the other-cell-to-same-cell interference ratio seen by the BS receiver and $\eta(t)$ is a parameter included in order to limit the total power received at the BS (Node B in UMTS). This parameter is conditioned by the most power demanding transmitting user. On the other hand, $C_{max}(t)$ could be considered as a measure of the maximum available capacity in the cell.

If P_i is the received signal power from user i and I_{total} the total received power, including noise power, in the BS, we can define $P_i = C_i \cdot I_{total}$, where C_i is the load factor of the user i. Considering the activity factor of user i, v_i, its average load factor is usually defined as L_i:

$$L_i = \frac{1}{\left(\dfrac{W}{v_i \cdot r_{i,k} \cdot \gamma_i}+1\right)} \qquad 3)$$

In the particular case of UMTS, provision of QoS in the UTRA FDD air interface is obviously related to functionalities of the protocol stack architecture of the interface, seen in 0. Physical layer [6] offers data transmission services to Medium Access Control (MAC) layer by means of Transport Channels (TrCh) [7]. The set of specific attributes of this layer (channel coding, interleaving, and transmission rate), which determines the transmission quality for the data information to be sent, is referred to as the Transport Format (TF) of the considered TrCh. The WCDMA air interface offers several options of transmitting packets using different dedicated, shared or common packet channels [7]. The MAC entity [8] is responsible for mapping Logical Channels (LCh) onto Transport Channels, selection of TF, priority handling and dynamic scheduling. A LCh is defined by the type of transferred information. In addition to physical procedures, upper layers provide other feasibilities depending on the service requirements. In particular, the Packet Data Convergence Protocol (PDPC) [9] contains compression methods, which are needed to get better spectral efficiency for services requiring IP packets to be transmitted over the radio interface. Whereas, the Radio Link Control (RLC) [10] protocol provides segmentation and reliable transmission services for both user and control data. This reliability is provided by means of Forward Error Correction (FEC) and Automatic Repeat reQuest (ARQ) techniques. To reduce the Bit Error Rate (BER), the first one uses coding data by adding overhead, whereas the second one uses retransmissions of erroneous packets, thus increasing delay. Another important feature of the RLC mode is the use of early discard, which allows in the transmitter to drop packets that have exceeded the maximum tolerable delay, reducing the delay of the following packets.

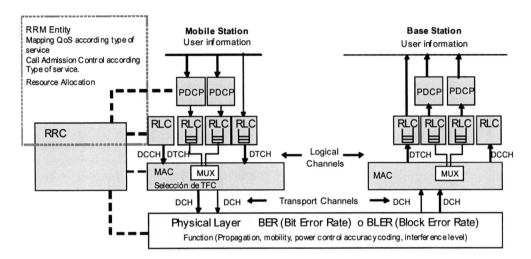

UTRA-FDD radio interface protocol architecture.

For successful transmission it is necessary that Radio Bearer Control selects the appropriated channels to match the QoS requirements. Common and shared channels are suitable for bursty traffic. Random Access CHannel (RACH) is typically used for transmission of short infrequent packets, while for larger amounts it is more appropriated to use a dedicated transport channel (DCH). On the other hand, in the last one, data transfer is more reliable due to the performed closed loop power control and the absence of collisions.

Within the UTRA protocol architecture, the set of RRM management task is associated with a set of control duty performed by the Radio Resource Control (RRC) entity [11] in order to support gathering information and communication among the involved elements in Radio Access Bearer QoS provision. Once the Radio Bearer Service (RB) is established, it is responsibility of Radio Resource Control (RRC) to ensure that traffic permitted to enter the network is within the limits of the admission region. Dynamic radio bearer control is performed by RRC, which configures the characteristics of the lower layer protocol entities in the air interface, including parameters for the Physical, Transport and Logical channels, based on the traffic volume measurements reported by Medium Access Control. At the same time it offers services to higher layers for signaling mobility management, call control and session management.

The current standard (Releases 99/R4/R5) supports both circuit and packet switched services using fixed and variable data rates. Dedicated channels are usually allocated for real-time applications, such as conversational and streaming classes in a circuit switched form. However, as the use of IP based services becomes more important, it is desirable to improve the performance on the uplink dedicated transport channels should be improved, allowing higher data rates for background, interactive and streaming based traffic transmitted in a packet switched form. Thus, we have limited the scope of the study to packet transmission over enhanced proposals of dedicated channels (Releases R6/R7) [12].

Current UMTS R99/R4/R5 DCH specifications support autonomous mobile station transmission and selection of TF among a TF Combination Set (TFCS) established and managed per MS, by RRC. TFCS reconfiguration is centralized in the RNC (Radio Network Controller), thus, latency and update rate are restricted by the communication delay between the RNC and Node B. We can say that, in the current specifications, the uplink scheduling and data rate control really reside in the RNC, which is not able to respond to fast changes in the uplink load.

One of the techniques being currently considered for enhanced uplink is Node B (BS) uplink scheduling control. Controlling the time of MS transmissions reduces latencies in rate control, exploits fast channel quality variations and consequently allows a most efficient and correct control of uplink interference levels. This control maximizes the throughput and allows supporting higher data rates. The method itself does not require changes in the DCH, but introduces new signaling to facilitate fast UL scheduling by means of transport format combination control. The scheduling of MS transmissions can be performed in two ways. Pure rate scheduling: all uplink transmissions occur in parallel, but at a low enough rate in order to satisfy the power control constraints and pure time scheduling: where only a subset of the MSs that have traffic to send is allowed to transmit at any given time, guaranteeing the minimum power control criterion

Node B scheduling, which has been adopted and improved in several previous works, [13][14], through the consideration of a centralized demand assignment protocol in the Medium Access Control (MAC) level, has been shown to provide differentiated QoS and an efficient use of the uplink power resources of the cell. Once the dedicated channel (code) is assigned, a MS needs to wait for the BS (Node B) to specify the Transport Format (TF), the power and the times in which it can transmit. Only a subset of the MSs that have traffic to send is allowed to transmit at any given time according to the pure time approach. Taking into account condition 2), scheduling at the BS arranges transmissions of MSs according to algorithms related with delay and rate requirements. MSs that cannot meet E_b/N_o requirements are delayed. Unused resources (power and instant of transmission) are assigned to the rest of the users. In addition to the scheduling strategies based on delay and rate requirements (round robin, Delay EDD, WFQ, etc), it is interesting to consider channel–state dependent scheduling algorithms based on the MS required value of parameter η. The ratio between inter and intracell interference, $f(t)$, and the channel condition should be known in order to achieve the minimum transmitted power criterion with accuracy. Note that, a priori, every mobile having permit to transmit must reach its E_b/N_o constraints. Only imperfections in the estimation of ratio $f(t)$ and in the channel estimation could prevent it.

Call Admission Control

Many of the proposed CAC policies for wired and wireless systems make admission decisions by comparing the resources required by an incoming connection request with the resources currently available in the network or with the scheduling region limits, expressed in terms of the subset of the space of the number of calls for which the packet level QoS (i.e:

BER, BLER, delay and jitter delay) constraints are met. However, this procedure implicitly assumes that the network capacity remains constant over the time, that is, the network has a fixed capacity. This can be more or less the case of wired or FDMA/TDMA wireless networks. In these networks, research efforts concentrate in low computational packet level characterization, GoS optimization (new call blocking probability and probability of dropping a call due to handoff), and frequency planning. In CDMA, although the need for frequency planning is eliminated, CAC becomes a more complex problem, since capacity is soft. In this case, the number of connections cannot longer specify a direct measure of the available capacity. The acceptance of a new user connection must be conditioned by the fact that signal to interference ratio, E_b/N_o , values can be achieved by each existing connection once a new one is activated. Ideally, call admission control should be able to accept a call only if a new equilibrium of the power control can be reached.

Most of the studies to determine an admission threshold are based on the maintenance of the quality of the signal in terms of SIR. Consideration of a packet based transmission mode requires to incorporate delay, delay jitter and packet loss probability as QoS parameters in the analysis of the capacity. This necessarily implies improving the link level behaviour characterization, since most of the proposals only consider physical level transmission and, in the best cases, consider a real implementation of the power control. Note that if the problem of the evaluation of the amount of resources needed by a variable rate connection is not simple in both wired and TDMA/FDMA wireless systems, in CDMA it is even more complicated, since the network resources used by a connection do not only depends on the bandwidth or/and the delay requirements, but also on the generated interference. In this context, results obtained from the packet level should be taken into account in order to infer and support an efficient method for CAC.

It is evident that to design an effective CAC has associated a first main design problem: to set an effective CAC threshold. This obviously makes practical capacity analysis for CDMA systems a challenging issue, especially in the presence of channel fading, imperfect power control, user movement and heterogeneous services. Next section presents a review of the proposed schemes for this CAC threshold. On the other hand, preventing forced termination of handoff calls and limiting the risk of blocking new calls is the other key requirement. It is known that, in order to increase the capacity, micro of picocell technology is widely used in cellular communications networks and particularly in wireless IP networks. Frequent handovers introduce a new paradigm in the area of network congestion and admission control. Mobile users may change their Base Station server a number of times during the lifetime of their call or connection; as a result, availability of wireless network resources could not be guaranteed for the whole call duration. On the other hand, this will result in rapidly changing traffic load in the network and frequent invocation of resource reservation mechanisms. Static and dynamic prioritization and sharing schemes, associated with static or adaptive reserves of resources (trunk reservation), could be proposed in order to allow the system capacity to be shared among the various traffic types and/or to protect the handoff connections from the new connections.

Soft Handoff

Handoff is the essential functionality for dealing with the mobility of the mobile stations (MSs). An advantage of CDMA is the viability of Soft Handoff (SHO). In SHO the user can get two or more simultaneous connections with different base stations (Nodes B). These base stations constitute the *active set*. Compared with the conventional Hard Handoff (HHO), SHO has the advantage of smoother transmission and less ping-pong effects. Although soft handoff has been shown to increase system capacity since it reduces interference from users in soft handoff operation, it requires a more complex control of MS transmissions in the packet level in order to guarantee the QoS requirements. Given that more than one BS control the cells where the MS is present, several alternatives need to be evaluated for the location of the scheduling entity that controls the MS transmissions: 1) To select one BS in the active set as scheduling entity for the MS (the best downlink BS or the best uplink BS can be chosen); 2) To consider multiple BS as scheduling entities. In this last case, each Node B in the active set generates scheduling commands taking into account its own available capacity. To consider a single Node B as scheduling entity has several advantages: it requires a smaller amount of downlink resources, because only one BS needs to send scheduling related information to a MS. Furthermore, the BS scheduler load is decreased because of the reduced number of MSs having to be scheduled per cell and, on the other hand, a MS needs to monitor the scheduling command only from the single scheduling BS. On the contrary, the accuracy of the uplink interference control may not be sufficient, since there is no scheduling coordination between the cells in SHO. Thus, scheduling BSs may cause significant amount of unexpected interference in other BSs included in the active set.

Given that the main objective is to extend the capacity on the uplink, a good power adjustment is required in order to optimize the radio resource management. The more promising approach is to assume that more than one cell could be considered as valid scheduling entities and to use the information obtained from the coordination between the cells in SHO in order to mitigate interference.

Several generic proposals have been considered by 3GPP with multiple scheduling Nodes B, which could have different scheduling aggressiveness [12]: 1) The MS follows the best scheduling command (maximum allowed rate). In this case all the other Nodes B may suffer from unexpected noise rise. 2) The MS follows the worst scheduling command (the maximum allowed rate is the minimum among the allowed rates signaled by the BS). In this case, the received quality would be better than expectation. 3) The MS can combine the scheduling commands by applying different weighting factors for each scheduling command. The network could determine the weighting factors. In this case, variations of the uplink interference from the neighbor cells are expected to be reduced.

One proposal slightly different from the previous references is presented in [15]. When a MS is allowed to transmit by more than one Node B, it selects the scheduling command with the maximum allowed data rate and the lowest power requirement from all the received scheduling commands.

This procedure ensures that the total uplink interference level in all the scheduling entities is almost always less than the level they have estimated in order to guarantee the required $(E_b/N_o)_{req,i,k}$ for all the scheduled MS's. Effects in the power control need to be computed in all the BS scheduling entities. Imperfections have also to be considered, in addition to modifications in the ARQ strategies performed in the RLC level.

Call Admission Control Threshold in CDMA Packet Based Wireless Networks

The problem of setting an effective CAC threshold has been addressed before in quite a lot of works, although in the most of them, radio resource and teletraffic management have been exclusively based on a single-service and/or consider the worst case scenario of a fixed, equal number or simply uniform distribution of calls in each cell. Moreover, there are few works that concern with exploiting the natural multiplexing gain and packet transmission approach in CDMA multimedia service systems and, usually, no specific MAC/Scheduling protocols are considered in the link level in order to provide differentiated QoS. This specific factor needs to be introduced in any effective capacity threshold analysis. However, results of most of these works combined with wired system experiences could be taken as a reference in order to infer a more efficient CAC.

In general, although setting a fixed threshold to CAC is hard, most of the research studies tried to obtain one in order to simplify the CAC decision. However, in these cases, CAC efficiency depends on a threshold that usually needs to be redesigned due to changes of the radio channel parameters and traffic statistics. On the other hand, from the implementation point of view, real time services usually require the use of a parameter-based CAC (PBAC), where user requirements and traffic performance are expressed as a function of parameters associated with a deterministic or stochastically model. However, the evaluation of the amount of resources needed by a set of variable rate connections in addition to the available capacity could be not simple. In this case, measurements performed over a period of time could provide an indicator of the system load and flow behavior. These approaches are similar to that of the measurement-based CAC (MBAC) proposed for wired networks, where decision is based on online measurements of the related traffic and resource statistics. Although in wired systems a clear differentiation exists between MBAC (more adaptable to the soft nature of multimedia systems) and PBAC approaches, in wireless networks a combination of PBAC and MBAC could be desirable.

Looking previous CAC research studies, the following approaches can be found in the most of these references:

- CAC that derives an average cell capacity threshold based on the allowed number of active connections.
- CAC based on the transmitted power limits, which prevents the power control to reach a new equilibrium in power assignment, on the total interference level or on the SIR guarantee.
- CAC based on throughput or on the "effective bandwidth "concept.

Beginning by the most traditional proposals, Viterbi, in [16], presents some expressions for system capacity evaluation that have been used by other authors given their calculus simplicity [17][18][19][20]. Call blocking probability is calculated from congestion probability and then Erlang capacity is calculated based on this blocking rate. In general terms, a new connection is blocked if total interference density at the base station (BS) exceeds the background noise level by an amount $1/\eta$ usually taken to be $10dB$. This criterion assures the receiver stability and assumes that the mean user bit error rate (BER) requirements could be maintained by a dynamic control of energy-to-interference (E_b/N_o) ratio. A distribution function of the interference is determined in order to evaluate the blocking probability. [16] shows that the mean and variance of other-cell interference can be approximated by the intracell interference, modeled as a Gaussian distribution, multiplied by a constant coefficient associated to the *ratio f*. However, since the Erlang capacity is calculated based only on the blocking rate, the instantaneous communication quality of the individual users is not guaranteed given that statistical multiplexing is not perfect and its relation with Erlang capacity is left unclear.

In a similar way the studies are conducted in [17][18][19]. An analogous procedure to [16] is applied in [18] to an integrated voice/data CDMA system, although in this case lognormal distribution is considered as a better alternative for intra-cell interference characterization. In [17] expressions of the QoS (packet loss) and GoS (grade of service) as functions of traffic intensity and CAC thresholds have been analytically derived, although only a type of service is considered. Unlike [16], inter-cell interference is expressed in terms of an *"equivalent"* number of users, whose probability density function (pdf) distribution is obtained from computer simulations. Gaussian and Gamma distributions are used as a model and contrary to [16], mean and standard deviation are approximated by the intra-cell interference (mean number of active users) multiplied by two different constant coefficients. [17] is probably the most representative reference of numbered based CAC schemes, although, in fact, two CAC thresholds are considered, in terms of the number of connections (NCAC) and in terms of the interference level (ICAC). When a NCAC threshold is considered, a fixed capacity model analogous to those for FDMA/TDMA systems is implicitly considered. Erlang B computes the blocking probability, whereas the quality loss is calculated as the probability that the number of active users overflows the system capacity. If an ICAC threshold is contemplated, the number of calls is not limited in advance. A new call is blocked if the observed interference level exceeds a predefined threshold T_{block}. Simulation results show that T_{block} is a more robust threshold that the number of users, given that it shows only slight variations when system parameters (path loss, shadowing, traffic's spatial distribution and transmission rate) change, whereas the NCAC threshold must be redesigned.

Interactive CAC [21] is based on the idea of admitting a new call if after an interactive admission phase, where quality of a new mobile could be allowed to be below its target, a new power control equilibrium can be reached in order to guarantee QoS [21]. The CAC policy can be considered ideal, since a new call is admitted if the system currently provides the required SIR level to all the calls. However, it presents serious convergence speed problems and it cannot exploit discontinuous transmission.

A signal to interference ratio (SIR)-based CAC algorithm is proposed in [22] and improved in [23][24][25][26]. A concept of residual capacity is introduced in [22] as the additional number of calls that BS can accept such loss probability could be guaranteed to remain below a certain level. The residual capacity is updated dynamically according to the reverse-link SIR measurements at the BS and the defined SIR threshold value. If the residual capacity is greater than zero, the system can accept a new connection. The greatest contribution of [22] is the assumption of non-uniform traffic distributions in cells and the consideration of CAC's that compute the effects that the new call admission has over adjacent cells in addition to the own cell effects. The assumption of a mono-service scenario facilitates the estimation of the thresholds. A priori, the adaptation to a multi-service scheme does not seem simple if the original format is maintained. On the other hand, SIR measures would not provide all the information necessary to incorporate in the analysis the improvements due to MAC or scheduling strategies. It is obvious that a new dimension should be added to the problem.

In the above references, the assumption of a natural statistic multiplexing without considering packet level access control, supposes that QoS, in terms of SIR ratio, is violated every time that the number of connections waiting for service exceeds the available capacity. When this happens, it is considered that all the packets are destroyed, causing a high packet loss probability. This assumption is clearly too conservative, since it is evident that the application of a flow control mechanism in the packet level should increase the system efficiency, limiting the number of simultaneously transmitting users. Moreover, most analysis concentrate simply on transmission characteristics and do not consider the real time operation of the system under stochastically varying traffic load. In [16][17][18] the ratio between inter-cell and intra-cell interference is considered fixed, although it is really variable and its fluctuations have an important influence over quality and capacity. In [16], the use of a tight upper bound value in order to characterize the stochastic nature of the ratio does not allow exploiting the advantages of statistical multiplexing. When adjacent cells are slight loaded, it is obvious that the reference cell could either increase transmission rates of previously delayed users or support a higher number of transmissions with QoS. Alternatively, if the peak value is replaced by the expected value, connections are not prevented against bad conditions for significant periods. Therefore, it is desirable an appropriate statistical or effective value characterization in order to exploit those advantages.

One possible approach to handle the randomness of both interference and traffic sources is to consider an effective capacity, lying between the mean and the peak value, in a similar way that the effective bandwidth concept associated to variable bit rate sources in call admission in the broadband integrated services digital network (ISDN) using asynchronous transfer mode (ATM) [27]. In the context of CDMA networks, there are few antecedents of the use of this concept. In the context of TDMA wireless networks, we can make reference to some studies, [28][29][30], that really combine the radio channel and traffic sources variability. Nevertheless, results of that context avoid a key requirement of CDMA: the correlation between the interference and the capacity.

Returning to the CDMA proposals, the concept is associated to different parameters. In [31], [32] and [20] this concept is applied to the effective bit rate, the signal to interference density ratio or the load factor, respectively. In [31], the stochastic nature of traffic, delay and BER requirements are used to characterize the resources required by multimedia services in a packet transmission system. The CAC threshold is expressed in terms of maximum number of allowed sources. From the equivalent capacity and the Gaussian approximation methods used in broadband wired networks [27], the bandwidth required by a single source to satisfy packet lost due to buffer overflow (buffer size is chosen in such a way that delay constraints are met) is calculated. Then, the ratio between the capacity of the BS wired output link to the backbone and the equivalent capacity provides a measure of the maximum number of allowed users. On the other hand, similar procedures to [16] are used to find the maximum number of users to provide wireless error rate guarantees. The proposed method, although seems to be useful, has been shown to be conservative. On the other hand, results depend on the selected value for the capacity of the BS wired output link and the relationship between this capacity and the one provided by radio access is unclear, since the last one really depends on the source QoS requirements.

In [32] the Gaussian approximation [27] and also the Chernoff bound are used to set an equivalent signal to interference density ratio to each traffic class as a measure of its effective bandwidth. The proposed model becomes similar to a fixed capacity system, with W (chip rate) the available capacity. Although calculus even involves CAC effects in both own and adjacent cells, the method does not exploit the multiplexing gain resulting from flow control and, in addition, the calculus of the effective bandwidth involves a high computational cost.

In [20] an average "*load factor*" associated to each service class is used in an analogous sense to the effective bandwidth. In accordance with 3), the load factor is defined as the ratio between the user received power and the total interference level. In order to exploit the advantages of variable capacity, it is proposed to adapt the CAC decision to the interference distribution, modeled in a similar sense that [16]. Loss probability and blocking and dropping probabilities are computed considering analogous expressions to [17]. The method however does not exploit the delay tolerance of some real time transmissions and does not guarantee a strict QoS in terms of packet loss. In fact, this proposal could be classified as throughput based CAC.

Effective bandwidth based CAC, introducing link QoS requirements: rate, delay and packet dropping probabilities, is developed in [33]. This work differs from the earlier ones in the following aspect: performance results obtained from packet level are used to infer an effective capacity threshold. Note that in any of the previous references, no specific MAC/Scheduling protocol has been developed in order to provide differentiated QoS. Considering the stochastic nature of multimedia traffic and the changes in the available capacity in the CDMA system, an effective capacity request, $C_{i,ef}$, is used to characterize the resources required by each mobile user in order to meet its respective QoS requirements. A new call is admitted if

$$\sum_{i \in cell} C_{i,ef} + C_{new,ef} \leq E[C_{max}] \qquad\qquad 4)$$

where $E[C_{max}]$ is the mean of the available capacity, as defined in 2). $C_{i,ef}$ can be calculated as 5), where $E[C_{max}(T)]$ is the mean of the available capacity, C_{max}, estimated in a period of time T, and N_{max} is the capacity of the system in terms of number of MSs that satisfy the required limit of packet dropping probability (ε_i). This requirement is related with the probability that the packet delay ($delay_{k,i}$) associated to a connection k belonging to class i exceeds its delay requirements ($D_{max,i}$). To obtain N_{max}, an analytical model is used in order to compute the delay distribution for individual sources considering their BLER and service rate requirements, the variable wireless capacity and the error control scheme (selective repeat algorithm is implemented) [33]. Stochastic changes in the available capacity are obtained both considering the stochastic nature of the ratio f in addition to changes in the own cell interference due to statistic changes on traffic source transmissions.

$$C_{i,ef} = E[C_{max}(T)] / N_{max} \qquad \text{5)}$$

$$N_{max} = max(\ Ni\ |\ \forall_{k=0:Ni}\ \ Pr[delay_{k,i} > D\,max_i] \le \varepsilon_i\ ) \qquad \text{6)}$$

Integration of several real time users with different BLER and rate requirements is immediate just considering the effective capacity computed separately for each type of traffic. So, a linear approximation can be used to find the total resources required by all the MSs. However, when different delay requirements are considered, it could be necessary, in some cases, to define a little reserve of capacity in order to prevent little deviations from the linear approximation. Really, the effective capacity depends on QoS and on the mix of other traffic sources. However, simulation results of the proposed method in [34] shows that, the most of the time, it is possible to consider conservative equivalent bandwidths, which can be shown to be independent on the mix of traffic sources and which suffer only very slight variations.

In the Interference-based CAC strategy proposed in [35], a new user is not admitted by the uplink CAC if the new resulting total interference level is higher than a threshold value, $I_{threshold}$, 7) (see 0 and expression 7)). This threshold is the same as the maximum uplink noise rise and can be set by radio network planning (usually limited to 10dB). The CAC algorithm also needs to estimate the increase level of interference increase, ΔI, associated to the new user by 8), where the increase of load associated to the new connection, ΔL, is in fact an effective capacity calculated considering 3).

$$I_{total\_old} + \Delta I \le I_{threshold} \qquad \text{7)}$$

$$\Delta I = \frac{I_{total\_old}}{1 - \eta_{UL} - \Delta L} \Delta L \qquad \text{with} \qquad \eta_{UL} = (1+f) \cdot \sum_{j \in Cellk} L_j \qquad \text{8)}$$

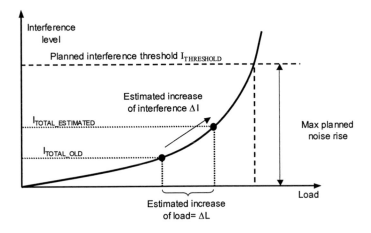

Uplink load curve and the estimation of the load increase due to a new user.

Although this threshold, coordinated with an appropriate flow control in the packet level, could guarantee the signal to interference ratio for all admitted connections, packet level quality in terms of BER is not assured within the delay constraints requirements unless the appropriate threshold value is recomputed by simulations and adapted in order to guarantee QoS [36]. However, the logic of this approach has made that other authors have taken this procedure as a reference, incorporating some improvements as the consideration of the interference increase in adjacent cells in addition to the own cell interference.

CAC approaches that consider effective bandwidth could be considered a good alternative in order to capture multiplexing gain, although the current proposed methods have limitations. Obtaining an accurate parameter can be more or less complicated but, once an effective capacity is derived for the variable rate connection, the behavior of the wireless multimedia system could be considered as a multirate circuit-switched or virtual circuit-switched system. On the other hand, although effective capacities associated to traffic sources need to be designed based on a stochastically and semi analytical model, the exact design is not always needed, and the CAC threshold designed for a uniform traffic distribution can be used in all cells for non-uniform traffic distributions, considering both homogeneous and heterogeneous load distributions in cells.

Motivations for using the concept of effective bandwidth are multiples: It allows QoS requirements associated to the connection to be mapped in a simple parameter; CAC becomes simple and effective; Techniques or policies of reserve of resources handle the capacity thresholds and the effective bandwidth estimations with greater facility; Network planning is simplified, if it is possible to resort techniques already used in networks ATM, such as the Roberts approach [37].

Mobility Support for QoS Provisioning

Provision of QoS in the call level focuses on developing call admission policies to minimize the forced termination of handoff calls and the risk of unnecessary blocking of new calls. Various handoff priority-based CAC schemes have been proposed. They can be classified into two broad categories:

In *guard channel schemes*, some resources are reserved for handoff calls. Reserve resources could be fixed or variable, although adaptive reserves, based on the estimated position, movement and the desired QoS for each class of traffic associated to mobiles stations are more desirable. On the other hand, in *fractional guard channel schemes* new calls are admitted with a certain probability, which depends on the number of busy resources. This scheme is more general than the first one.

There exist a second group of solutions, which permit queueing the handover calls and /or serving them with a higher priority than new call requests if a congested radio cell is encountered.

In wireless multimedia networks, the priority schemes implemented to protect horizontal or vertical handoff calls from the new ones should be applied to multimedia services (streaming, interactive,...) in order to provide a flexible tool for QoS differentiation.

There exist other solutions, which permit queueing the handover calls and /or serving them with higher priority relative to new call requests in a congested radio cell.

Reservation of resources and mobile's call arrival prediction are closed related. An accurate prediction of the mobile's path reduces the number of BS that may reserve resources for handover calls and consequently improves the overall system efficiency. Allocating resources in all future cells that the MS may visit for the time interval during which the MS will reside in each cell, is only feasible if an exact knowledge of the mobile path and arrival and departure times to every cell along the path is available. Obtaining an exact knowledge of MS mobility is not possible in most cases, due to the uncertainty of the mobile environments and the difficulty in specifying the mobility profiles of MS.

Mobility Modelling and Teletraffic Aspects

From the point of view of the mobility, the most interesting parameters are the channel holding time and the handoff probability. These parameters are necessary to obtain an accurate analysis of call blocking probability (P_{BLOCK}) and handoff call dropping probability (P_{LOSS}). The channel holding time is defined as the time spent in a cell by a user in communication prior to handoff (or subsequent handoff) or the time until the call completion. It is the minimum of two random variables associated with the call/connection holding time and the cell dwell time, respectively. The cell dwell time or the cell residence time is the time a MS spends within a BS coverage, so that a link of acceptable quality can be maintained. Depending on whether a call is originated in a cell or handoff from a neighbouring cell, two different cell residence times must be specified: the new call residence time, $\bar{\tau}_{dwell} = 1/\mu_{i,dwell}$, and the handover call cell residence time, $\bar{\tau}_{dwell\_sub} = 1/\mu_{i,dwell\_sub}$, respectively. Both times are two random variables whose distributions have to be found. A realistic characterization is important in order to have an appropriate traffic model that reflects the traffic situation and the user mobility patterns. For the sake of convenience and tractability, most traffic analysis make the assumption that call holding and dwell times are exponentially distributed and consequently the channel holding time, although the assumption could be not too realistic. Factors such as mobility and cell shape and size cause the dwell time to have a different probability distribution function to that of call duration. This difference could be greater for higher mobility and smaller cell sizes. Several authors suggest more realistic random variables in order to model new and handover call residence times. A gamma, lognormal or a mixture of lognormals, sum of hyper exponentials, or hyper_erlang distributions could be considered. However, in general the objective is to retain the Markovian properties that are required to model the call level performance by a multidimensional birth-death process.

Nevertheless, given that P_{BLOCK} and P_{LOSS} are insensitive to the channel holding time distribution, when the assumption of both new and handoff call arrival are Poisson distributed is true, we can adopt the exponential assumption. Analytical models will likely have to be coupled with measurement based mechanism in order to estimate mean dwell times for each traffic class according to its mobility pattern, given that handoff probabilities are easily derived from new and handoff channel holding times. Note that although effects of using exponential distribution in general network analysis could be relative, these effects will be higher if we plan to apply them to the estimation of the amount of resources we must reserve in a cell. In this case, obviously, a more accurate model is necessary. Reservation of resources and MS call arrival prediction are closely related. An accurate prediction of the mobile's path reduces the number of BS that may reserve resources for handover calls and consequently the overall system efficiency.

To obtain an analytical solution, under the assumption of Poisson arrivals, there are two basic approaches: the first one is based on multidimensional Markov chains, where the system states are described by the number of active calls of each traffic class. The multi-dimensional state space has as many dimensions as the number of traffic classes. In the absence of a product form, which is the case of considering adaptive resource reservation schemes, calculating the resource occupancy distribution involves to numerically solve the balance equations, which is prohibitively demanding. Alternatively, as a second approach, it is possible to use several methods for approximating the blocking probabilities. In [37] are considered and evaluated one-dimensional methods based on solving the marginal distribution of resource occupancy (*Roberts* and *Convolutional* approach); and two-dimensional approximations where the joint distribution of the pairs is considered: total amount of capacity in use and amount of capacity in use by each class of traffic sources. In the *Roberts* method, the multi-dimensional the state space is mapped into a one-dimensional state space without affecting the resulting blocking and handoff loss probabilities. The number of occupied basic resource units defines the states of the traffic model and a recursive solution is proposed in order to obtain GoS parameters. Each call class can require variable number of resources and the method can be extended to trunk/adaptive reservation schemes. Although a multi-dimensional approach is more accurate, in most cases of practical interest results provided by the simplest one-dimensional approach could be considered good enough. The analytical model through the Roberts approach has been shown effective in [38] in order to assess the performance of the system in terms of P_{BLOCK} and P_{LOSS}, assuming an "effective bandwidth" based threshold and an adaptive resource reservation scheme at the call level.

Further Discussion and Summary

The chapter presents a survey on the issues related to the adaptations of cellular access networks based on a CDMA access technique to the context of IP based 4G wireless networks. A collection of radio resource management schemes requires being adapted to packet based transmissions contemplated in 4G.

Assuming Diffserv as the most promising architecture to the global network, a desirable approach is to consider at least two independent domains with different conditions and requirements, one for the wired part and the other for the packet based evolution of the wireless part. Restricting the scope of the chapter to the specific CDMA wireless access, differentiated allocation of resources (power, rate, etc) through power control, prioritised access through packet scheduling, call admission control and performance guaranties during soft handoff have been discussed.

A cross layer design has been considered as a promising solution to meet the various QoS requirements of multimedia traffic over wireless links and to improve the overall system performance. Joint power and rate assignment, through centralized packet scheduling at the Base Station in coordination with an optimum power criterion, is one of the main subjects of RRM. Additionally, challenges in designing efficient CAC schemes adapted to the new 4G systems context have been analyzed and a comparison of several CAC thresholds for CDMA systems has been presented. In general, a CAC cross-layer design should be applied to capture both call and packet-level QoS performances. Teletraffic aspects related with the mobility have been outlined.

References

[1] R. Braden, D. Clark, and S. Shenker, "Integrated services in the Internet architecture: An overview", *IETF, RFC* **1663**, 1994.

[2] S. Blake et al., "An Architecture for Differentiated Services", *IETF RFC* **2475**, 1998.

[3] 3GPP Technical Specification 23.107 "Quality of Service (QoS) concept and architecture".

[4] 3GPP Technical Specification 23.207 "End-to-end Quality of Service (QoS) concept and architecture".

[5] A. Sampath, P. S. Kumar, J.M. Holtzman, "Power Control and Resource Management for a Multimedia Wireless CDMA System", *Proceedings IEEE PIMRC'95*, pp.21-25.

[6] 3GPP Technical Specification 25.214 "Physical layer procedures (FDD)".

[7] 3GPP Technical Specification 25.211 "Physical channels and mapping of transport channels onto physical channels (FDD)".

[8] 3GPP Technical Specification 25.321 "Medium Access Control (MAC) protocol specification".

[9] 3GPP Technical Specification 25.323 "Packet Data Convergence Protocol (PDCP) specification".

[10] 3GPP Technical Specification 25.322 "Radio Link Control (RLC) protocol specification".

[11] 3GPP Technical Specification 25.922 "Radio resource management strategies".

[12] 3GPP Technical Report 25.896, "Feasibility Study for Enhanced Uplink for UTRA FDD".

[13] A. Hernández, F. Casadevall, "Scheduling and Quality of Service in W-CDMA", *Proceedings AMOS ACTS MOBILE SUMMIT*, Sorrento (Italia), 1999, pp. 795-800.

[14] A. Hernández, A. Valdovinos, F. Casadevall, "Performance Analysis of Packet Scheduling Strategies for Multimedia Traffic in WCDMA", *Proceedings IEEE VTC'2002* Spring, 2002, pp. 155-159.

[15] A. Hernández, A. Valdovinos, "Scheduling and Call Admission Control Schemes in Soft Handoff for Packet Switched Transmission in WCDMA Networks", *Proceedings IEEE VTC'04 Fall*, 2004, pp.3486-3490.

[16] A. M. Viterbi, A. J. Viterbi, "Erlang capacity of a Power Controlled CDMA System", *IEEE Journal on Selected Areas in Communications*, vol. 11, no. 6, August. 1993, pp. 892-900.

[17] Y. Ishikawa, N. Umeda, "Capacity Design and Performance of Call Admission Control in Cellular CDMA Systems", *IEEE Journal on Selected Areas in Communications*, vol. 15, no. 8, October 1997, pp. 1627-35.

[18] A. Sampath, B. Mandayan, J. M. Holzman, "Erlang Capacity of Power Controlled Integrated Voice and Data CDMA System", *Proceedings IEEE VTC'97*.

[19] A. Sampath, J. M. Holtzmam, "Access Control of Data in Integrated Voice/Data CDMA Systems: Benefit and Tradeoffs", *IEEE Journal on Selected Areas in Communications*, vol. 15, no. 8, October 1997, pp. 1511-1526.

[20] V. Phan-Van, D. D. Luong, "Capacity Enhancement with Simple and Robust Soft-Decision Call Admission Control for WCDMA Mobile Cellular PCN's", *Proceedings IEEE VTC'2001*.

[21] D. Kim, "Efficient Interactive Call Admission Control in Power-Controlled Mobile Systems", *IEEE Transactions on Vehicular Technology*, vol. 49, no. 3, May 2000, pp. 1017-1028.

[22] Z. Liu, M. El Zarki, "SIR-Based Call Admission Control for DS-CDMA Cellular Systems", *IEEE Journal on Selected Areas in Communications.*, vol. 12, no. 4, 1994, pp. 638-644.

[23] M. Kim, B-C. Shin, D-J. Lee, "SIR-Based Call Admission Control by Intercell Interference Prediction for DS-CDMA Systems", *IEEE Communications letters*, vol. 4, no. 1, January 2000, pp. 29-31.

[24] D. Kim, "On Upper Bounds of SIR-Based Call Admission Threshold in Power-Controlled DS-CDMA Mobile Systems, *IEEE Communications letters*, vol. 6, no. 1, January 2002, pp. 13-15.

[25] M. H. Ahmed, H. Yanikomeroglu, "A Lower Bound on SIR Threshold of Call Admission Control in Multiple-Class CDMA Systems with Imperfect Power-Control," *Proceedings IEEE Globecom'04*.

[26] M. H. Ahmed, H. Yanikomeroglu, "SINR Threshold Lower Bound for SINR-based Call Admission Control in CDMA Networks with Imperfect Power Control" *IEEE Communications letters*, vol. 9, no. 4, 2005, pp. 1-3.

[27] M. Schwartz, "Broadband Integrated Networks", Ed. Prentice Hall, 1996.

[28] M. Krunz, J. G. Kim, "Fluid analysis of delay and packet discard performance for QoS support in wireless networks," *IEEE Journal on Selected. Areas in Communications*, vol. 19, no. 2, Feb. 2001, pp. 384-395.

[29] D. Wu, R. Negi, "Effective Capacity: A wireless Link Model for Support of Quality of Service", *IEEE Transactions on Wireless Communications*, vol. 2, no. 4, July 2003, pp 630-643.

[30] M. Hassan; M.M. Krunz, I. Matta, "Markov-based channel characterization for tractable performance analysis in wireless packet networks", *IEEE Transactions on Wireless Communications*, vol. 3, May 2004, pp. 821–831.

[31] J.Q. J. Chak, W. Zhuang, "Capacity Analysis for Connection Admission Control in Indoor Multimedia CDMA Wireless Communications", *Wireless Personal Communications* no. 12, 2000, pp. 269-282.

[32] J.S. Evans, D. Everitt, "Effective Bandwidth- Based Admission Control for Multiservice CDMA Cellular Networks", *IEEE Transactions on Vehicular Technology*, vol. 48, no. 1, January 1999, pp. 36-46.

[33] A. Hernández, A. Valdovinos, F. Casadevall. "Capacity Analysis and Call Admission Techniques for CDMA Packet Transmission Systems", *Proceedings IEEE MWCN'02*, 2002, pp. 355-359.

[34] A. Hernández, A. Valdovinos, F. Casadevall. "Capacity Analysis and Performance Evaluation of Call Admission Control for Multimedia Packet Transmission in UMTS WCDMA System" *Proceedings IEEE WCNC 2003*, 2003, pp. 1550-1555.

[35] H. Holma, A. Toskala, "WCDMA for UMTS. Radio Access For Third Generation Mobile Communications", John Wiley &Sons, 2000.

[36] A. Hernández, A. Valdovinos, "Threshold Schemes Comparison for Call Admission Control in 3G Packet-Based CDMA Wireless Networks". *Proceedings WPMC'04*, 2004, pp. 421-426.

[37] S. C. Borst, D. Mitra, "Virtual Partitioning for Robust Resource Sharing Computational Techniques for Heterogeneous Traffic", *IEEE Journal on Selected Areas in Communications*, vol. 16, no 5, June 1998, pp. 668-678.

[38] A. Hernández, A. Valdovinos, F. Casadevall. "QoS and Radio Resource Management in Multimedia Packet Transmission for 3G Wireless IP Networks" *Proceedings IEEE VTC'04* Spring, 2004, pp. 1983-1987.

In: Mobile Multimedia: Communication Engineering ...   ISBN 1-60021-207-7
Editor: I.K. Ibrahim and D. Taniar pp. 259-278   © 2006 Nova Science Publishers, Inc.

Chapter 13

ADVANCED POWER REDUCTION TECHNIQUES IN MOBILE COMPUTING SYSTEMS

Ben Abdallah Abderazek, and Masahiro Sowa* [†]
National University of Electro-communications, Graduate School of
Information Systems, 1828585 Chofu-shi, Tokyo

Abstract

In recent years, the rapid expansion of wireless services is an indication that significant value is placed on portability and accessibility as key features of mobile communication. Mobile devices have maximum utility when they can be used anywhere and anytime. One of the major limitations of that goal is limited power supply. Since batteries provide limited power, a general constraint of mobile communication is the short continuous operation time of mobile devices. Therefore, power reduction and management is one of the most challenging problems in mobile systems.

Energy conservation has been largely considered in the hardware design of the mobile terminal and in components such as CPU, disk, display, memory, etc. Significant additional power savings can be also achieved by incorporating low-power methods into the design of network protocols used for data (audio, video, etc.) communication.

This chapter investigates in details power reduction techniques at the mobile host's components level, such as CPU, disk drive, screen back light, etc., and at the network protocols level used for data communication.

PACS 05.45-a, 52.35.Mw, 96.50.Fm. **Key Words**: mobile computing, power optimizations techniques, energy, multimedia.

AMS Subject Classification: 53D, 37C, 65P.

1   Introduction

Computation and communication have been steadily moving toward mobile and portable devices. With continued miniaturization and increasing computation power, we see ever

---

[*]E-mail address: ben@is.uec.ac.jp
[†]E-mail address: sowa@is.uec.ac.jp

growing use of powerful microprocessors running sophisticated, intelligent control software in a vast array of devices including pagers, cellular phones, laptop computers, digital cameras, video cameras, video games, etc. Unfortunately, there is an inherent conflict in the design goals behind these devices: as mobile systems, they should be designed to maximize battery life, but as intelligent devices, they need powerful processors, which consume more energy than those in simpler devices, thus reducing battery life.

In spite of continuous advances in semiconductor and battery technologies that allow microprocessors to provide much greater computation per unit of energy and longer total battery life, the fundamental tradeoffs between performance and battery life remains critically important [12, 13, 14, 40].

Multimedia applications and mobile computing are two trends that have a new application domain and market. Personal mobile or ubiquitous computing is playing a significant role in driving technology. An important issue for these devices will be the user interface-the interaction with its owner. The device needs to support multimedia tasks and handles many different classes of data traffic over a limited bandwidth wireless connection, including delay sensitive, real-time traffic such as video and speech.

Wireless networking greatly enhances the utility of a personal computing device. It provides mobile users with versatile communication, and permits continuous access to services and resources of the land-based network. A wireless infrastructure capable of supporting packet data and multimedia services in addition to voice will bootstrap on the success of the Internet, and in turn drive novel networked applications and services. However, the technological challenges to establishing this paradigm of personal mobile computing are non-trivial. In particular, these devices have limited battery resources. While reduction of the physical dimensions of batteries is a promising solution, such effort alone will reduce the amount of charge retained by the batteries. This will in turn reduce the amount of time a user can use the computing device. Such restrictions tend to undermine the notion of mobile computing. In addition, more extensive and continuous use of network services will only aggravate this problem since communication consumes relatively much energy. Unfortunately, the rate at which battery performance improves is very slow, despite the great interest created by the wireless business.

The energy efficiency is an issue involving all layers of the system, its physical layer, its communication protocol stack, its system architecture, its operating system, and the entire network [14]. This implicates several mechanisms that can be used to attain a high-energy efficiency. There are several motivations for energy-efficient design. Perhaps the most visible driving source is the success and growth of the portable consumer electronic market.

In its most abstract form, a networked system has two sources of energy drain required for its operation: (1) Communication, due to energy spent by the wireless interface and due to the internal traffic between various parts of the system, and (2) Computation, due to processing for applications, tasks required during communication, and operating system. Thus, minimizing energy consumption is a task that will require minimizing the contributions of communication and computation.

From another had, power consumption has become a major concern because of the ever-increasing density of solid-state electronic devices, coupled with an increasing use of mobile computers and portable communication devices. The technology has thus far helped to build low power systems. The speed-power efficiency has indeed gone up since 1990 by 10 times each 2.5 years for general-purpose processors and digital signal processors (DSPs) [13].

Design for low-energy consumption is certainly not a new research field, and yet remains one of the most difficult as future mobile system designers attempt to pack more capabilities such as multimedia processing and high bandwidth radios into battery operated portable miniature packages. Playing times of only a few hours for personal audio, notebooks, and cordless phones are clearly not very consumer friendly. Also, the required batteries are voluminous and heavy, often leading to bulky and unappealing products [15].

The key to energy efficiency in future mobile systems will be, then, designing higher layers of the mobile system, their functionality, their system architecture, their operating system, and the entire network, with energy efficiency in mind.

In this chapter we address a variety of energy reduction techniques that can be used for building an energy-efficient system. In section 2 , we present design optmizations techniques at the technology-level. Section three presents the power aware design techniques at the system-kevel. Section 4 gives power reduction design techniques at the system level. Finally, conclusion is given in section 5.

2   Power Aware Technological-Level Design Optimizations

2.1   Factors Affecting CMOS Power Consumption

Most components in a mobile system are currently fabricated using CMOS technology. Since CMOS circuits do not dissipate power if they are not switching, a major focus of low power design is to reduce the switching activity to the minimal level required to perform the computations [28, 29].

The sources of energy consumption on a CMOS chip can be classified as static and dynamic power dissipation. The average power is given by:

$$P_{avg} = P_{static} + P_{dynamic} \tag{1}$$

The static power consumption is given by:

$$P_{static} = P_{short-ciruit} + P_{leak} = I_{sc}.V_{dd} + I_{leak}.V_{dd} \tag{2}$$

and the dynamic power consumption is given by:

$$P_{dynamic} =_{\alpha_0 \to 1} C_L.V_{dd}^2.f_{clk} \tag{3}$$

The three major sources of power dissipation are, then, summarized in the following equation:

$$P_{avg} =_{\alpha_0 \to 1} C_L.V_{dd}^2.f_{clk} + I_{sc}.V_{dd} + I_{leak}.V_{dd} \tag{4}$$

The first term of formula 4, represents the switching component of power, where $\alpha_{0 \to 1}$ is the node transition activity factor (the average number of times the node makes a power consuming transition in one clock period), C_L is the load capacitance and f_{clk} is the clock frequency. The second term is due to the direct-path short circuit current, I_{sc}, which arises when both the NMOS and PMOS transistors are simultaneously active, conducting current directly from supply ground. The last term, I_{leak} (leakage current), which can arise from substrate injection and subthreshold effects, is primarily determined by fabrication technology.

$\alpha_{0 \to 1}$ is defined as the average number of times in each clock cycle that a node with capacitance, C_L, will make a power consuming transition resulting in an average switching component of power for a CMOS gate to be simplified to:

$$P_{switch} = \alpha_{0 \to 1} \; C_L . V_{dd}^2 . f_{clk} \tag{5}$$

Since the energy expended for each switching event in CMOS circuits is $C_L . V_{dd}^2 . f_{clk}$, it has the extremely important characteristics that it becomes quadratically more efficient as the high transition voltage level is reduced.

It is clear that operating at the lowest possible voltage is most desirable, however, this comes at the cost of increased delays and thus reduced throughput. It is also possible to reduce the power by choosing an architecture that minimizes the effective switched capacitance at a fixed voltage: through reductions in the number of operations, the interconnect capacitance, internal bit widths and using operations that require less energy per computation. We will use Formula (4) and (5) to discuss the energy reduction techniques and trade-offs that involve energy consumption of digital circuits. From these formula, we can see that there are four ways to reduce power: (1) reduce the capacity laod C, (2) reduce the supply voltage V, (3) reduce the switching frequency f, and (4) reduce the switching activity.

2.2   Reducing Voltage and Frequency

Supply voltage scaling has been the most adopted approach to power optimization, since it normally yields considerable savings thanks to the quadratic dependence of P_{switch} on V_{dd} [28]. The major shortcoming of this solution, however, is that lowering the supply voltage affects circuit speed. As a consequence, both design and technological solutions must be applied in order to compensate the decrease in circuit performance introduced by reduced voltage. In other words, speed optimization is applied first, followed by supply voltage scaling, which brings the design back to its original timing, but with a lower power requirement.

It is well known that reducing clock frequency f alone does not reduce energy, since to do the same work the system must run longer. As the voltage is reduced, the delay increases. A common approach to power reduction is to first increase the speed performance of the module itself, followed by supply voltage scaling, which brings the design back to its original timing, but with a lower power requirements [29].

A similar problem, i.e., performance decrease, is encountered when power optimization is obtained through frequency scaling. Techniques that rely on reductions of the clock frequency to lower power consumption are thus usable under the constraint that some performance slack does exist. Although this may seldom occur for designs considered in their entirety, it happens quite often that some specific units in a larger architecture do not need peak performance for some clock/machine cycles. Selective frequency scaling (as well as voltage scaling) on such units may thus be applied, at no penalty in the overall system speed.

2.3   Reducing Capacitance

Energy consumption in CMOS circuitry is proportional to capacitance C. Therefore, a path that can be followed to reduce energy consumption is to minimize the capacitance. A significant fraction of a CMOS chips energy consumption is often contributed to driving large off-chip capacitances, and not to core processing. Off-chip capacitances are in the order of five to tens of picofarads. For conventional packaging technologies, pins contribute approximately 13-14 pF of capacitance each (10 pF for the pad and 3-4 pF for the printed circuit board) [32].

From our earlier discussion, equation (5) indicates that energy consumption is proportional to capacitance, I/O power can be a significant portion of the overall energy consumption of the chip. Therefore, in order to save energy, use few external outputs, and have them switch as infrequently as possible. Packaging technology can have a impact on the energy consumption. For example, in multi-chip modules where all of the chips of a system are mounted on a single substrate and placed in a single package, the capacitance is reduced. Also, accessing external memory consumes much energy. So, a way to reduce capacitance is to reduce external accesses and optimise the system by using on-chip resources like caches and registers.

2.3.1   Chip Layout

There are a number of layout-level techniques that can be applied. Since the physical capacitance of the higher metal layers are smaller, there is some advantage to select upper level metals to route high-activity signals. Furthermore, traditional placement involves reducing area and delay, which in turn translates to minimizing the physical capacitance of wires. Placement that incorporates energy consumption, concentrates on minimizing the activity-capacitance product rather than capacitance alone. In general, high-activity wires should be kept short and local. Tools have been developed that use this basic strategy to achieve about 18% reduction in energy consumption.

The capacitance is an important factor for the energy consumption of a system. However, reducing the capacity is not the distinctive feature of low-power design, since in CMOS technology energy is consumed only when the capacitance is switched. It is more important to concentrate on the switching activity and the number of signals that need to be switched. Architectural design decisions have more impact than solely reducing the capacitance.

2.3.2   Technology Scaling

Scaling advanced CMOS technology to the next generation improves performance, increases transistor density, and reduces power consumption. Technology scaling typically has three main goals: (1) reduce gate delay by 30%, resulting in an increase in operating frequency of about 43%; (2) double transistor density; and (3) reduce energy per transistor by about 65%, saving 50% of the power. These are not ad hoc goals; rather, they follow scaling theory [32].

As the Semiconductor Industry Association roadmap (SIA) indicates, the trend of process technology improvement is expected to continue for years [31]. Scaling of the physical dimension involves reducing all dimensions: thus transistor widths and lengths are reduced, interconnection length is reduced, etc. Consequently, the delay, capacitance and energy consumption will decrease substantially.

Another way to reduce capacitance at the technology level is to reduce chip area. For example, an energy efficient architecture that occupies a larger area can reduce the overall energy consumption, e.g. by exploiting locality in a parallel implementation.

3　Power Aware Logic-Level Design Optimizations

Logic-level power optimization has been extensively researched in the last few years. While most traditional power optimization techniques for logic cells focus on minimizing switching power, circuit design for leakage power reduction is also gaining importance [33]. As a result, logic-level design can have a high impact on the energy-efficiency and performance of the system. Issues in the logic level relate to for example state-machines, clock gating, encoding, and the use of parallel architectures.

3.1　Clock Gating

Several power minimization techniques work especially well at the logic level. Most of them rely on switching frequency. The best example of which is the use of clock gating [16]. Clock gating provides a way to selectively stop the clock, and thus force the original circuit to make no transition, whenever the computation to be carried out by a hardware unit at the next clock cycle is useless. In other words, the clock signal is disabled to shut down some modules of the chip, that are inactive. This saves on clock power, because the local clock line is not toggling all the time.

For example the latency for the CPU of the TMS320C5x DSP processor [34] to return to active operation from the IDLE3 mode takes around $50\mu s$, due to the need of the on-chip PLL circuit to lock with the external clock generator. With the conventional scheme, the

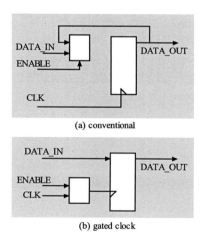

Figure 1: Clock gating example

register is clocked all the time, whether new data is to be captured or not. If the register must hold the old state, its output is fed back into the data input through a multiplexer whose enable line (ENABLE) controls whether the register clocks in new data or recycles the existing data. However, with a gated clock, the signal that would otherwise control the select line on the multiplexer now controls the gate. The result is that the energy consumed in driving the registers clock input (CLK) is reduced in proportion to the decrease in average local clock frequency. The two circuits function identically, but utilization of the gated clock reduces the power consumption.

3.2   Logic Encoding

The power consumption can be also reduced by carefully minimizing the number of transitions. The designer of a digital circuit often has the freedom of choosing the encoding scheme. Different encoding implementations often lead to different area, power, and delay trade-offs. An appropriate choice of the representation of the signals can have a big impact on the switching activity.

The frequency of consecutive patterns in the traffic streams is the basis for the effectiveness of encoding mechanisms. For example, a program counter in a processor generally uses a binary code. On average two bits are changed for each state transition [35]. Using a Gray-code (single bit changes) can give interesting energy savings. However, a Gray-code incremental requires more transistors to implement than a ripple carry incrementer [35]. Therefore, a combination can be used in which only the most frequently changing LSB bits use a Gray code.

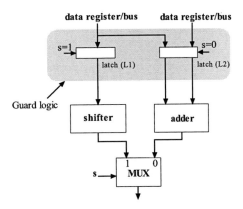

Figure 2: Dual Operation ALU with Guard Logic. The multiplexer does the selection only after both units have completed their evaluation. The evaluation of one of the two units is avoided by using a guard-logic; two latches (L1 and L2) are placed with enable signals (s1 and s2) at the inputs of the shifter and the adder respectively.

3.3   Data Guarding

Switching activity is the major cause of energy dissipation in most CMOS digital systems. Therefore, to reduce power consumption, switching activities that do not contribute to the actual communication and computation should be eliminated. The basic idea is to identify logical conditions at some inputs to a logic circuit that is invariant to the output. Since those input values do not affect the output, the input transitions can be disabled.

Data logic-guarding technique [36], is an efficient method used to guard not useful switching activities to propagate further inside the system. The technique is based on reducing the switching activities by placing transparent latches/registers with an enable signal at the inputs of each block of the circuit that needs to be selectively turned off. If the module is to be active in a clock cycle, the enable signal makes the latch transparent, permitting normal operation. If not, the latch retains its previous state and no transitions propagate through the inactive module (see Figure 2). As a summary, the logic-level design can have a high impact on the energy-efficiency and the performance of a given system. Even with the use of state of the arts hardware design language (i.e., Verilog HDL), there are still many optimizations techniques that should be explored by the designers to reduce the energy consumption at the logic-level. The most effective technique used at this level is the reduction of switching activities.

4   Power-Aware System Level Design Optimizations

In the previous sections we have explored sources of energy consumption and showed the low level - technology and circuit levels, design techniques used to reduce the power dissipation. In this section, we will concentrate on the energy reduction techniques at the architecture and system level.

4.1   Hardware System Architecture Power Consumption Optimizations

The implementation dependent part of the power consumption of a system is strongly related to the number of properties that a given system or algorithm may have. The component that contributes a significant amount of the total energy consumption is the communication channel or interconnect.

Experiments have already been made in designs and proved that about 10 to 40% of the total power may be dissipated in buses, multiplexers and drivers [7, 37]. This amount can increase dramatically for systems with multiple chips due to large off-chip bus capacitance. The energy consumption of the communication channels is largely dependent on algorithm and architecture-level design decisions. Regularity and locality are two important properties of algorithms and architectures for reducing the energy consumption due to the communication channels. The idea behind regularity is to capture the degree to which common patterns appear in an algorithm. Common patterns enable the design of less complex architecture and therefore simpler interconnect structure and less control hardware. Simple measures of regularity include the number of loops in the algorithm and the ratio of operations to nodes in the data flow graph. The statistics of the percentage of operations covered by sets of patterns is also indicative of an algorithm's regularity. Quantifying this measure involves first finding a promising set of patterns, large patterns being favoured. The core idea is to grow pairs of as large as possible isomorphic regions from corresponding pairs of seed nodes [38].

Locality relates to the degree to which a system or algorithm has natural isolated clusters of operation or storage with few interconnections between them. Partitioning the system or algorithm into spatially local clusters ensures that the majority of the data transfers take place within the clusters and relatively few between clusters. The result is that the local buses with a low electrical capacity are shorter and more frequently used than the longer highly capacitive global buses. Locality of reference can be used to partition memories. Current high-level synthesis tools are targeted to area minimization or performance optimization. However, for power reduction it is better to reduce the number of accesses to long global buses and have the local buses be accessed more frequently. In a direct implementation targeted at area optimization, hardware sharing between operations might occur, destroying the locality of computation. An architecture and implementation should preserve the locality and partition and implement it such that hardware sharing is limited. The increase in the number of functional units does not necessarily translate into a corresponding increase in the overall area and energy consumption since the localization of interconnect allows a more compact layout and also fewer access to buffers and multiplexers are needed.

4.1.1   Hierarchical Memory System

Efficient use of an optimised custom memory hierarchy to exploit temporal locality in the data accesses can have a very large impact on the power consumption in data dominated applications.   The idea of using a custom memory hierarchy to minimize the power consumption is based on the fact that memory power consumption depends primarily on the access frequency and the size of the memory. For on-chip memories memory power increases with the memory size. In practice, the relation is between linear and logarithmic depending on the memory library. For off chip memories, the power is much less dependent on the size because they are internally heavily partitioned. Still they consume more energy per access than the smaller on-chip memories. Hence, power savings can be obtained by accessing heavily used data from smaller memories instead of from large background memories [21, 22].

As most of the time only a small memory is read, the energy consumption is reduced. Memory considerations must also be taken into account in the design of any system. By employing an on-chip cache significant power reductions together with a performance increase can be gained. Apart from caching data and instructions at the hardware level, caching is also applied in the file system of an operating system [21]. The larger the cache, the better performance. Energy consumption is reduced because data is kept locally, and thus requires less data traffic. Furthermore, the energy consumption is reduced because less disk and network activity is required.

The compiler also has impact on power consumption by reducing the number of instructions with memory operands. It also can generate code that exploits the characteristics of the machine and avoids expensive stalls. The most energy can be saved by a proper utilization of registers. In [39], a detailed review of some compiler techniques that are of interest in the power minimization arena is also presented.

Secondary Storage

Secondary storage in modern mobile systems generally consists of a magnetic disk supplemented by a small amount of DRAM used as a disk cache; this cache may be in the CPU main memory, the disk controller, or both [24, 25, 26]. Such a cache improves the overall performance of secondary storage. It also reduces its power consumption by reducing the load on the hard disk, which consumes more power than the DRAM.

Energy consumption is reduced because data is kept locally, and thus requires less data traffic. In addition, the energy consumption is reduced because less disk and network traffic is required. Unfortunately, there is trade-off in size of the cache memory since the required amount of additional DRAM can use as much as energy as a conventional spinning hard disk [23].

A possible technology for secondary storage is an integrated circuit called flash memory [26]. Like a hard disk, such memory is non-volatile and can hold data without consuming energy. Furthermore, when reading or writing, it consumes only 0.15 to 0.47 W, far less than a hard disk. It has a read speed of about 85 ns per byte, quite like DRAM, but write speed of about $410\mu s$ per byte, about 10 to 100 times slower than hard disk. However, since flash memory has no seek time, its overall write performance is not that much worse than a magnetic disk; in fact, for sufficiently small random writes, it can actually be faster. Since flash is practically as fast as DRAM at reads, a disk cache in no longer important for read operation. The cost per megabyte of flash is about 7 to 40 times more expensive than guard disk, but about 2 to 5 times less expensive than DRAM. Thus, flash memory might also be effective as a second level cache bellow the standard DRAM disk cache [26, 24].

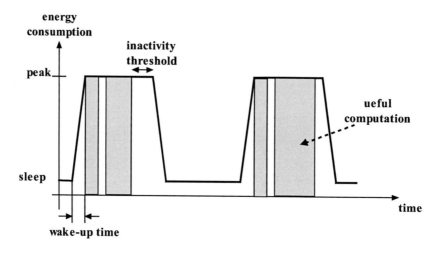

Figure 3: Power consumption in typical processor

4.1.2   Processor

In general, the power consumption of the CPU is related to the clock rate, the supply voltage, and the capacitance of the devices being switched [16, 18, 19, 20]. One power-saving feature is the ability to slow down the clock. Another is the ability to selectively shut off functional units, such as the floating-point unit; this ability is generally not externally controllable. Such a unit is usually turned off by stopping the clock propagated to it. Finally, there is the ability to shut down processor operation altogether so that it consumes little or no energy. When this last ability is used, the processor typically returns to full power when the next interrupt occurs. A time energy consumption relation ships is given in Figure 3. Turning off a processor has little downside; no excess energy is expended turning the processor back on, the time until it comes back on is barely noticeable, and the state of the processor is unchanged from it turning off and on, unless it has a volatile cache [16]. Therefore, reducing the power consumption of the processor can have a greater effect on overall power savings than it might seem from merely examining the percentage of total power attributable to the processor.

4.1.3   Display and Backlight

The display and backlight have very few energy-saving features. This is unfortunate, since they consume a great deal of power in their maximum-power states; for instance, on the Duo 280c, the display consumes a maximum of 0.75 W and the backlight consumes a maximum of 3.40 [11, 40]. The backlight can have its power reduced by reducing the brightness level or by turning it off, since its power consumption is roughly proportional to the luminance delivered. The display power consumption can be reduced by turning the display off. It can also be reduced slightly by switching from colour to monochrome or by reducing the update frequency, which reduces the range of shades or colours of Gray for each pixel, since such shading is done by electrically selecting each pixel for a particular fraction of its duty cycle. Generally, the only disadvantage of these low-power modes is reduced readability. However, in the case of switches among update frequencies and switches between colour and monochrome, the transitions can also cause annoying flashes.

4.2   Operating System Power Consumption Optimization

Software and algorithmic considerations can also have a severe impact on energy consumption[2, 7, 8, 10, 39, 40]. Digital hardware designers have promptly reacted to the challenge posed by low-power design. Designer skills, technology improvements and CAD tools have been successful in reducing the energy consumption. Unfortunately, software engineers and system architects are often less "energy-aware" than digital designers, and they also lack suitable tools to estimate the energy consumption of their designs. As a result, energy-efficient hardware is often employed in a way that does not make optimal use of energy saving possibilities. In this section we will show several approaches to reduce energy consumption at the operating system level and to the applications. A fundamental OS task is efficient management of host resources. With energy as the focus, the question becomes how to make the basic interactions of hardware and software as energy efficient as possible for local computation. One issue observed in traditional performance-centric resource management involves latency hiding techniques. A significant difference and challenge in energy-centric resource management is that power consumption is not easy to hide.

As one instance of power-aware resource management, we consider memory management. Memory instructions are among the more power-hungry operations on embedded processors [8], making the hardware/software of memory management a good candidate for optimization. Intels guidelines for mobile power [9, 17] indicate that the target for main memory should be approximately 4% of the power budget. This percentage can dramatically increase in systems with low power processors, displays, or without hard disks. Since

Table 1: Operating system functionality and corresponding techniques for optimizing energy utilization.

CPU scheduling	Idle power mode, voltage scaling
Operating system functionality	Energy efficient techniques
Memory allocation	Adaptive placement of memory blocks, switching of hardware energy reduction modes
Application/OS interaction	Agile content negotiation trading fidelity for power, APIs
Resource Protection and allocation	Fair distribution of battery life among both local and distributed tasks, lockingbatery for expensive operations
Communication	Adaptive network polling, energy-aware routing, placement of distributed computation, and server binding

many small devices have no secondary storage and rely on memory to retain data, there are power costs for memory even in otherwise idle systems. The amount of memory available in mobile devices is expanding with each new model to support more demanding applications (i.e., multimedia) while the demand for longer battery life also continues to grow significantly.

Scheduling is needed in a system when multiple functional units need to access the same object. In operating systems scheduling is applied at several parts of a system for processor time, communication, disk access, etc. Currently scheduling is performed on criteria like priority, latency, time requirements etc. Power consumption is in general only a minor criterion for scheduling, despite the fact that much energy could be saved.

Subsystems of a computer, such as the CPU, the communication device, and storage system have small usage duty cycles. That is, they are often idle and wait for the user or network interaction. Furthermore, they have huge differences in energy consumption between their operating states.

Recent advances in ad hoc networks allow mobile devices to communicate with one another, even in the absence of pre-existing base-stations or routers. All mobile devices are able to act as routers, forwarding packets among devices that may otherwise be out of communication range of one another. Important challenges include discovering and evaluating available routes among mobile devices and maintaining these routes as devices move, continuously changing the topology of the underlying wireless network. In applications with limited battery power, it is important to minimize energy consumption in supporting this ad-hoc communication.

There are numerous opportunities for power optimizations in such environments, including: i) reducing transmission power adaptively based on the distance between sender and receiver, ii) adaptively setting transmission power in route discovery protocols, iii) balancing hop count and latency against power consumption in choosing the best route between two hosts, and iv) choosing routes to fairly distribute the routing duties (and the associated power consumption) among nodes in an ad-hoc network [27].

4.3   Application, Compilation Techniques and Algorithm

In traditional power-managed systems, the hardware attempts to provide automatic power management in a way that is transparent to the applications and users. This has resulted in some legendary user problems such as screens going blank during video or slide-show presentations, annoying delays while disks spin up unexpectedly, and low battery life because of inappropriate device usage. Because the applications have direct knowledge of how the user is using the system to perform some function, this knowledge must penetrate into the power management decision-making system in order to prevent the kinds of user problems described above. This suggests that operating systems ought to provide application programming interfaces so that energy-aware applications may influence the scheduling of the systems resources.

The switching activity in a circuit is also a function of the present inputs and the previous state of the circuit. Thus it is expected that the energy consumed during execution of a particular instruction will vary depending on what the previous instruction was. Thus an appropriate reordering of instructions in a program can result in lower energy. Today, the cost function in most compilers is either speed or code size, so the most straightforward way to proceed is to modify the objective function used by existing code obtimizers to obtain low-power versions of a given software program. The energy cost of each instruction must be considered during code obtimization. An energy aware compiler has to make a trade-off between size and speed in favour of energy reduction.

At the algorithm level functional pipelining, retiming, algebraic transformations and loop transformations can be used [8]. The system's essential power dissipation can be estimated by a weighted sum of the number of operations in the algorithm that has to be performed. The weights used for the different operations should reflect the respective capacitance switched. The size and the complexity of an algorithm (e.g. operation counts, word length) determine the activity. Operand reduction includes common sub-expression elimination, dead code elimination etc. Strength reduction can be applied to replace energy consuming operations by a combination of simpler operations (for example by replacing multiplications into shift and add operations).

4.4   Energy Reduction in Network Protocols

Up to this point we have mainly discussed the techniques that can be used to decrease the energy consumption of digital systems and focused on the computing components of a mobile host. In this subsection we will discuss some techniques that can be used to reduce the energy consumption that is needed for the communication external of the mobile host. We classify the sources of power consumption, with regard to network operations, into two types: (1) communication related and (2) computation related.

Communication involves usage of the transceiver at the source, intermediate (in the case of ad hoc networks), and destination nodes. The transmitter is used for sending control, route request and response, as well as data packets originating at or routed through the transmitting node. The receiver is used to receive data and control packets some of which are destined for the receiving node and some of which are forwarded. Understanding the power characteristics of the mobile radio used in wireless devices is important for the efficient design of communication protocols.

The computation mainly involves usage of the CPU, main memory, the storage device and other components. Also, data compression techniques, which reduce packet length, may result in increased power consumption due to increased computation. There exists a potential trade-off between computation and communication costs. Techniques that strive to achieve lower communication costs may result in higher computation needs, and vice-versa. Hence, protocols that are developed with energy efficiency goals should attempt to strike a balance between the two costs.

Energy reduction should be considered in the whole system of the mobile and through all layers of the protocol stack. The following discussion presents some general guidelines that may be adopted for an energy efficient protocol design.

4.4.1   Protocol Stack Energy Reduction

Data communication protocols dictate the way in which electronic devices and systems exchange information by specifying a set of rules that should a consistent, regular, and well-understood data transfer service. Mobile systems have strict constraints on the energy consumption, the communication bandwidth available, and are required to handle many classes of data transfer over a limited bandwidth wireless connection, including real time traffic such as speed and video. For example, multimedia applications are characterized by their various media streams with different quality of service requirements.

In order to save energy an obvious mode of operation of the mobile host will be a sleep mode [6]. To support such mode the network protocols need to be modified. Store-and-forward schemes for wireless networks, such as the IEEE 802.11 proposed sleep mode, not only allow a network interface to enter a sleep mode but can also perform local retransmissions not involving the higher network protocol layers.

There are several techniques used to reduce the power consumption in all layers within the protocol stack. In Figure 4, we list areas in which conservation mechanisms are efficient.

Collisions should be eliminated as much as possible within the media access layer (MAC) layer, a sublayer of the data link layer, since they result in retransmissions. Retransmissions lead to unnecessary power consumption and to possibly unbounded delays. Retransmissions cannot be completely avoided in a wireless network due to the high error-rates. Similarly, it may not be possible to fully eliminate collisions in a wireless mobile network. This is partly due to user mobility and a constantly varying set of mobiles in a cell.

For example, new users registering with the base station may have to use some form of random access protocol. In this case, using a small packet size for registration and bandwidth request may reduce energy consumption. The EC-MAC protocol [6] is one example that avoids collisions during reservation and data packet transmission. This is the default mechanism used in the IEEE 802.11 wireless protocol in which the receiver is expected to keep track of channel status through constant monitoring. One solution is to broadcast a schedule that contains data transmission starting times for each mobile as in [6]. Another solution is to turn off the transceiver whenever the node determines that it will not be receiving data for a period of time.

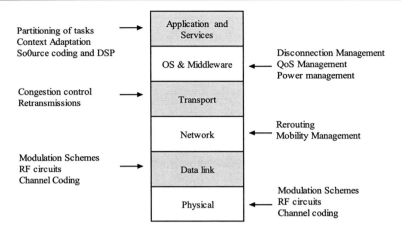

Figure 4: Protocol stack of a generic wireless network, and corresponding areas of energy efficient possible research.

Physical layer:

As shown in Figure 4, the lowest level of the protocol stack is the physical layer. This layer consists of radio frequency (RF) circuits, modulation, and channel coding systems. At this level, we need to use an energy-efficient radio that can be in various operating modes (like variable RF power and different sleep modes) such that it allows a dynamic power management [5]. Energy can also be saved if it is able to adapt its modulation techniques and basic error-correction schemes. The energy per bit transmitted or received tends to be lower at higher bit rates. For example, the WaveLAN radio operates at 2Mb/s and consumes 1.8 W, or 0.9 J/bit. A commercially available FM transceiver (Radiometrix BIM-433) operates at 40 kb/s and consumes 60 mW, or 1.5 J/bit. This makes the low bit-rate radio less efficient in energy consumption for the same amount of data. However, when a mobile has to listen for a longer period for a broadcast or wake-up from the base station, then the high bit-rate radio consumes about 30 times more energy than the low bit rate radio. Therefore, the low bit-rate radio must be used for the basic signalling only, and as little as possible for data transfer. To minimise the energy consumption, but also to mitigate interference and increase network capacity, the transmit power on the link should be minimised, if possible.

Data link Layer:

The data link layer is thus responsible for wireless link error control, security (encryption/decryption),mapping network layer packets into frames, and packet retransmission. A sublayer of the data link layer, the media access control (MAC) protocol layer is responsible for allocating the time-frequency or code space among mobiles sharing wireless channels in a region.

In an energy efficient MAC protocol the basic objective is to minimise all actions of the network interface, i.e. minimise on-time of the transmitter as well as the receiver. Another way to reduce energy consumption is by minimising the number of transitions the wireless interface has to make. By scheduling data transfers in bulk, an inactive terminal is allowed to doze and power off the receiver as long as the network interface is reactivated at the scheduled time to transceive the data at full speed.

An example of an energy-efficient MAC protocol is E^2MaC [4]. The E2MaC protocol is designed to provide QoS to various service classes with a low energy consumption of the mobile. In this protocol, the main complexity is moved from the mobile to the base station with plenty of energy. The scheduler of the base station is responsible to provide the connections on the wireless link the required QoS and tries to minimise the amount of energy spend by the mobile. The main principles of the E^2MaC protocol are avoid unsuccessful actions, minimise the number of transitions, and synchronise the mobile and the base-station.

Network layer:

The network layer is responsible for routing packets, establishing the network service type, and transferring packets between the transport and link layers. In a mobile environment this layer has the added responsibility of rerouting packets and mobility management. Errors on the wireless link can be propagated in the protocol stack. In the presence of a high packet error rate and periods of intermittent connectivity of wireless links, some network protocols (such as TCP) may overreact to packet losses, mistaking them for congestion. TCP responds to all losses by invoking congestion control and avoidance algorithms. These measures result in an unnecessary reduction in the link's bandwidth utilisation and increases in energy consumption because it leads to a longer transfer time.

The limitations of TCP can be overcome by a more adequate congestion control during packet errors. These schemes choose from a variety of mechanisms to improve end-to-end throughput, such as local retransmissions, split connections and forward error correction. A comparative analysis of several techniques to improve the end-to-end performance of TCP over lossy, wireless hops is given [3]. These schemes are classified into three categories: end-to-end protocols, where loss recovery is performed by the sender; link-layer protocols, that provide local reliability; and split-connection protocols that break the end-to-end connection into two parts at the base station. The results show that a reliable link-layer protocol with some knowledge of TCP provides good performance, more than using a split-connection approach. Selective acknowledgement schemes are useful, especially when the losses occur in bursts.

OS and middleware layer:

The operating system and middleware layer handles disconnection, adaptively support, and power and QoS management within wireless devices. This is in addition to the conventional tasks such as process scheduling and file system management. To avoid the high cost, in terms of performance, energy consumption or money, of wireless network communication is to avoid use of the network when it is expensive by predicting future access and fetching necessary data when the network is cheap. In the higher level protocols of a communication system caching and scheduling can be used to control the transmission of messages. This works in particular well when the computer system has the ability to use various networking infrastructures (depending on the availability of the infrastructure at a certain locality), with varying and multiple network connectivity and with different characteristics and costs. True prescience, of course, requires knowledge of the future. Two possible techniques, LRU caching and hoarding, are for example present in the Coda cache manager. A summary of of other software strategies for energy efficiency is presented in [1, 2].

5   Conclusion

This chapter has investigated a number of energy-aware design techniques in a mobile system. In particular, it covered techniques used to design energy-aware systems at the technology, logic, and system levels. The vast majority of the techniques used at the system architectures, are derived from existing uni-processor energy-aware mobile systems.
Other issue to be considered but not discussed in this chapter are the inextricable link between energy-aware design techniques and their impact on mobile system reliability.

Acknowledgments

The authors would like to thank all anonymous reviewers for providing several useful comments for this work.

References

[1] J. Kistler, Disconnected operation in a distributed file system, PhD thesis, Carnegie Mellon University, School of Computer Science,(1993).

[2] J.R. Lorch and A.J. Smith, Software strategies for portable computer energy management, *IEEE Personal Communications* **5**(3) 6073,(1998).

[3] H.Balakrishnan, V.enkata N. Padmanabhan, A Comparison of Mechanisms for Improving TCP Performance over Wireless Links, *IEEE/ACM Transaction on Networking*, Vol. 5, No. 6,pp.756-769,(1997).

[4] P. J. Havinga, G. Smit, M. Bos M., *Energy-efficient wireless ATM design*,proceedings wmATM99, June 2-4, (1999).

[5] F. Akyildiz, S. Weilian , S. Yogesh, and E. Cayirci, A Survey on Sensor Networks, *IEEE Communications Magazine*, pp.102-114, (2002).

[6] K.M. Sivalingam, J.C. Chen, P. Agrawal and M. Srivastava, Design and analysis of low-power access protocols for wireless and mobile ATM networks, *Wireless Networks* **6**(1), 7387, (2000) .

[7] J. Liang, et. al., An architecture and compiler for scalable on-chip communication. *IEEE Trans. on VLSI Systems*, **12**(7), (2004).

[8] V. Tiwari, S.Malik, and A. Wolfe. Power analysis of embedded software: A first step towards software power minimization, *IEEE Transactions on Very Large Scale Integration*, **2**(4):437445,(1994).

[9] Intel Corporation. Mobile Power Guidelines 2000. ftp://download.intel.com/design/mobile/ intelpower/mpg99r1.pdf, December 1998.

[10] F. Wolf, Behavioral Intervals in Embedded Software: Timing and Power Analysis of Embedded Real-Time Software Process, Kluwer Academic Publishers, ISBN 1-4020-7135-3, 2002

[11] J. Lorch, A complete picture of the energy consumption of a portable computer. Master's thesis, Department of Computer Science, University of California at Berkeley,(1995).

[12] T. Martin, Balancing batteries, power and performance: System issues in CPU speed-setting for mobile computing,Ph.D. Dissertation, Carnegie Mellon University, Department of Electrical and Computer Engineering, Aug. (1999).

[13] Gregory F. Welch, A Survey of Power Management Techniques in Mobile Computing Operating Systems, *ACM SIGOPS Operating Systems Review*, Volume 29, Issue 4, (1995).

[14] R. Kravets and P. Krishnan, Application driven power management for mobile communication Springer Science,*Wireless Networks,*Vol.6, No. 4, pp. 263-277, (2000).

[15] J.M. Rulnick and N. Bambos, Mobile power management for maximum battery life in wireless communication networks, in: *Proceedings of IEEE INFOCOM 96,* (1996).

[16] L. Benini and G. de Micheli,System-level power optimization: Techniques and tools, in *Proc. Int. Symp. Low-Power Electronics Design*, San Diego, CA, pp. 288293, (1999).

[17] Intel Corporation. *Mobile Intel Pentium III processor in BGA2 and micro-PGA2 packages*, revision 7.0, (2001).

[18] J. R. Lorch and A. J. Smith. Reducing processor power consumption by improving processor time management in a single user operating system. In *Second ACM International Conference on Mobile Computing and Networking* (MOBICOM),(1996).

[19] M. Weiser, B. Welch, A. Demers, and S. Shenker. Schedlibng for reduced cpu energy. In Proceedings of the *First Symposium on Operating System Design and Implementation* (OSDI) 94, (1994).

[20] K. Govil, E. Chan, and H. Wasserman. Comparing algorithms for dynamic speed-setting of a low-power cpu, In *First ACM International Conference on Mobile Computing and Networking* (MOBICOM), (1995).

[21] C.L. Su and Alvin M. Despain, Cache Designs for Energy Efficiency, in *Proc. of the 28th Hawaii International Conference on System Science*, (1995).

[22] C. Su and A. Despain, Cache Design Tradeoffs for Power and Performance Optimization: A Case Study, *Proc. of International Symposium on Low Power Design*, (1995).

[23] P. Erik, P. Harris, W. Steven , E. W. Pence, S. Kirkpatrick, Technology directions for portable computers. *Proceedings of the IEEE*, **83**(4):636657, (1995).

[24] F. Doughs, P. Krishnan, and B. Marsh, Thwarting the power hungry disk. In *Proceedings of the 1991 Winter USENIX Conference*, (1994).

[25] K. Li, R. Kumpf, P. Horton, and T. Anderson, A quantitative analysis of disk drive power manage ment in portable computers. In *Proceedings of the 1994 Winter USENIX*, (1994).

[26] F. Douglis, F. Kaashoek, B. March, R. Caceres, K. Li, and J. Tauber, storage alternative for mobile computers, *Proceedings of the first USENIX Symposimum on Operating Systems Design and IMplemnetation*, (1994).

[27] P. J. M. Havinga and G. J. M. Smit, Energy-efficient wireless networking for multimedia applications, in *Wireless Communications and Mobile Computing*. New York: Wiley,vol. 1, pp.165184, (2001).

[28] F. N. Najm, A survey of power estimation techniques in VLSI circuits, *IEEE Trans. VLSI Syst.*, vol. 2, no. 4, pp. 446455, (1994).

[29] M. Pedram, Power minimization in IC design: Principles and applications, *ACM Trans. Design Automat. Electron. Syst.*, vol. 1, no. 1, pp. 356, (1996).

[30] Multiprocessor System-on-Chip,Morgan Kaufman Pblishers,ISBN:0-12385-251-X, (2005).

[31] Semiconductor Industry Association:" The national technology rodmap for semiconductors: technology needs", Sematche Inc., htt://www.sematech.org, Austin, USA, (1997).

[32] S. Borkar, Design challenges of technology scaling. *IEEE Micro*, **19**(4), (1999).

[33] Y. Ye, S. Borkar, and V. De, A New Technique for Standby Leakage Reduction in High-Performance Circuits, 1998 *Symposium on VLSI Circuits,* pp. 4041, Honolulu, Hawaii, (1998)

[34] L. Benini, G. de Micheli, and E. Macii, Designing low-power circuits: Practical recipes,*IEEE Circuits Syst. Mag.*, vol. 1, pp. 625, (2001).

[35] B. A. Abderazek, Sotaro Kawata, Tsutomu Yoshinaga, and Masahiro Sowa, Modular Design Structure and High-Level Prototyping for Novel Embedded Processor Core. Proceedings of the 2005 IFIP International Conference on Embedded And Ubiquitous Computing (EUC'2005), Nagasaki, Dec. 6 - 9, pp. 340-349,(2005).

[36] V. Tiwari, S. Malik, and P. Ashar. Guarded Evaluation: Pushing Power Management to Logic Synthesis/Design. *IEEE Transactions on Computer Aided Design of Integrated Circuits and Systems*, **17**(10):10511060, (1998).

[37] A. Abnous, J. Rabaey, Ultra-Low-Power Domain-Specific Multimedia Processors, *Proceedings of the IEEE VLSI Signal Processing Workshop*, IEEE press, pp. 459-464,(1996).

[38] J. Rabaey, L. Guerra,R.Mehra, Design guidance in the Power Dimension, *Proceedings of the ICASSP*, (1995).

[39] H. Mehta, R. M. Owens, M. J. Irwin, R. Chen, and D. Ghosh, Techniques for Low Energy Software, in *Internatzonal Symposzum of Low Power Electronics and Deszgn*, pp. 72-75, IEEE/ACM, (1997).

[40] J. Lorch, Modeling the effect of different processor cycling techniques on power consumption,Performance Evaluation Group Technical Note 179, *ATG Integrated Sys., Apple Computer*, (1995).

In: Mobile Multimedia: Communication Engineering ...   ISBN: 1-60021-207-7
Editors: I.K. Ibrahim and D. Taniar pp. 279-295   © 2006 Nova Science Publishers, Inc.

Chapter 14

3G M-HEALTH SYSTEM PERFORMANCE

Eduardo Antonio Viruete Navarro[*], *José Ruiz Mas*[**] *and Julián Fernández Navajas*[1]

Communication Technologies Group – Aragon Institute of Engineering Research,
University of Zaragoza, 50018 Zaragoza, Spain

Abstract

The IPv4 performance of a multi-collaborative wireless telemedicine system operating over Third-Generation (3G) cellular networks is presented. The system is designed to communicate the personnel of an ambulance with medical specialists in a remote hospital through a Universal Mobile Telecommunication System (UMTS) wireless access. Its architecture is based on advanced signalling protocols that allow multimedia multi-collaborative conferences in IPv4/IPv6 3G scenarios. The system offers simultaneous transmission of real-time medical data and videoconference, together with other non real-time services. IPv4 performance results show that the system performs reliably over IPv4-only UMTS accesses (64 Kbps in the uplink). Bandwidth use and jitter buffer design and trade-offs are the main aspects considered. Measurements allow dimensioning system parameters in order to improve transmission efficiency, channel utilization and, finally, the quality of the services offered.

Introduction

Mobile Health (m-Health) is an emerging area of telemedicine in which the recent developments in mobile networks and telemedicine applications converge.   m-Health involves the exploitation of mobile telecommunication and multimedia technologies and their integration into new mobile healthcare delivery systems [1]. Wireless and mobile networks have brought about new possibilities in the field of telemedicine thanks to the wide coverage provided by cellular networks and the possibility of serving moving vehicles.

---

[*] E-mail address: eviruete@unizar.es
[**] E-mail address: jruiz@unizar.es
[1] E-mail address: navajas@unizar.es

One of the first wireless telemedical systems that utilized Second-Generation (2G) Global System for Mobile Communications (GSM) networks addressed the Electrocardiogram (ECG) transmission issues [2]. In recent years, several m-Health and wireless telemedical systems based on GSM have been reported [3], allowing the accomplishment of remote diagnosis in mobile environments, as well as communication to geographic zones inaccessible by wired networks. The recent developments in digital mobile telephonic technologies and their impact on mobility issues in different telemedical and telecare applications are clearly reflected in the fast growing commercial domain of mobile telemedical services. A comprehensive review of wireless telemedicine applications and the most recent advances on m-Health systems are presented in [4].

However, 2G-based systems lack the necessary bandwidth to transmit bandwidth-demanding medical data. The Third-Generation (3G) Universal Mobile Telecommunications System (UMTS) overcomes the limitations of first and second mobile network generations supporting a large variety of services with different Quality of Service (QoS) requirements. However, this fact makes network design and management much more complex. New applications require networks to be able to handle services with variable traffic conditions keeping the efficiency in the network resources utilization. The UMTS air interface is able to cope with variable and asymmetric bit rates, up to 2 Mbps and 384 kbps in indoor and outdoor environments, respectively, with different QoS requirements such as multimedia services with bandwidth on demand [5]. In this kind of scenario, the emergence of 3G mobile wireless networks will permit to extend the use of m-Health applications thanks to their higher transmission rates and flexibility over previous mobile technologies [6].

UMTS introduces the IP Multimedia core network Subsystem (IMS) [7], an IPv6 network domain designed to provide appropriate support for real-time multimedia services, independence from the access technologies and flexibility via the separation of access, transport and control. The fundamental reason for using IPv6 is the exhaustion of IPv4 addresses. Support for IPv4 is optional, but since network components require backward compatibility, it is clear that a dual stack configuration (IPv4 and IPv6) must be provided. The IMS uses the Session Initiation Protocol (SIP) as signalling and session control protocol [8]. SIP allows operators to integrate real-time multimedia services over multiple access technologies such as General Packet Radio Service (GPRS), UMTS or, ultimately, other wireless or even fixed network technologies (interworking multimedia domains).

This chapter presents the IPv4 performance of a 3G m-Health system (Fig. 1) [9] designed for different critical and emergency medical scenarios. With this system, medical specialists in a hospital take part in a multipoint conference with the medical personnel of an ambulance, receiving compressed and coded biomedical information from the patient, making it possible for them to assist in the diagnosis prior to his reception. The m-Health system includes intelligent modules such as information compression and coding, and QoS control (data prioritization, congestion control and jitter control) [10] to significantly improve the transmission efficiency of joint real-time and non real-time data over wireless channels in a more appropriate way than previous systems [11]. IPv4 performance results validate the QoS mechanisms used and allow to dimension jitter buffers in order to improve the quality of real-time services.

m-Health System Overview

The wireless m-Health system has been built using standard off-the-shelf hardware, instead of developing propriety hardware as in [12], uses free software and commercially available 3G wireless UMTS cellular data services. In the first stages of its design, user requirements and functional specifications were established in collaboration with medical specialists, in order to create a portable and modular system that could be easily integrated in any environment, using any underlying network technology capable of supporting IP multimedia services.

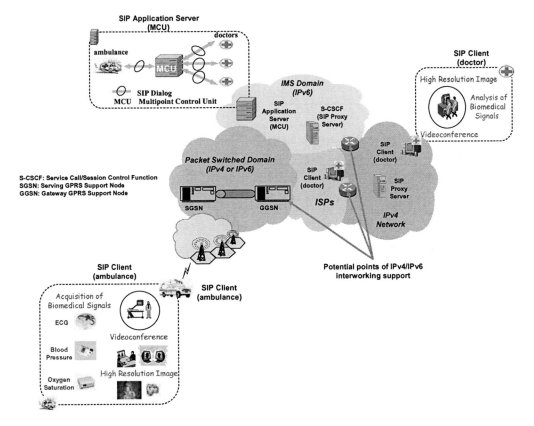

Figure 1: Wireless m-Health system

The details of the 3G m-Health system architecture [9] are shown in Fig. 1. The system comprises of the signalling and session control, medical user services and application control subsystems, together with the QoS control internal subsystem. Several intelligent modules allow the acquisition, treatment, representation and simultaneous media transmission, rather than only one media at a time [13]-[14]. Information compression, coding and QoS control (data prioritization, congestion control and jitter control) modules [10] improve the transmission efficiency of joint real-time and non real-time data over wireless channels in a more appropriate way than previous systems [11]. In addition, and also unlike [11], this system follows a multi-collaborative design, integrates new real-time multimedia features intended for 3G wireless networks, supports IPv4/IPv6 interworking [15] and uses SIP as the service control protocol, including messages defined specifically for the IMS by the 3rd Generation Partnership Project (3GPP). The IPv4/IPv6 SIP dual stack is the basis to integrate the ambulance and the hospital in any possible 3G scenario [16] (Fig. 2): IPv4 ambulance connecting to an IPv6 hospital, IPv6 communication through IPv4 islands and, finally, native all-IPv6 communication.

In the first stages of IPv6 deployment in 3G networks, there will be GPRS networks that will only be able to provide IPv4 connections. As it is shown in Fig. 2a, if an IPv4 ambulance needs to communicate with an IPv6 hospital, a transition mechanism is required. One of the possibilities is the use of a Network Address Protocol Translator-Port Translator (NAPT-PT) working together with a dual stack Domain Name System (DNS) server with a DNS-Application Level Gateway (DNS-ALG). The main problem with NAPT-PT is that it modifies packets, therefore security cannot be guaranteed if an IP Security (IPsec) association is used. But this is exactly the same problem that exists today with the use of IPv4 private addressing for peer-to-peer applications. On the other hand, the NAPT-PT mechanism provides an interoperability framework for more than one user terminal and any kind of application.

The next step of IPv6 deployment will be a common scenario when operators start providing IPv6 communications and customers start connecting to IPv6 services. Some IPv6 islands will still need to be connected through an IPv4 network (Fig. 2b) as it is done today in fixed networks. In this case, the most appropriate transition mechanism is tunneling. Tunneling means adding an IPv4 header instead of translating it, which could cause an important bandwidth waste in the Radio Access Network (RAN). However, this tunneling will occur in the fixed world.

Finally, the expected final scenario in IPv6 deployment (Fig. 2c) does not require any transition mechanism provided that the m-Health system is IPv6 capable and ready to be integrated in an all-IPv6 3G world.

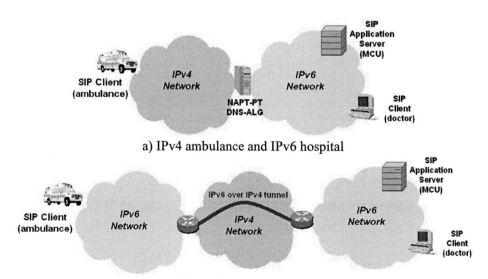

a) IPv4 ambulance and IPv6 hospital

b) IPv4 islands between IPv6 networks

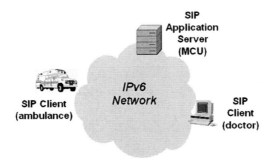

c) Native IPv6 communication

Figure 2: IPv4/IPv6 migration process

Communication between the remote medical personnel and medical specialists is established by means of multipoint multi-collaborative sessions through several network environments capable of supporting the different types of multimedia traffic. The conference model selected is the tightly coupled conference model [17], which requires the existence of a Multipoint Control Unit (MCU) (Fig. 1). System users and the MCU exchange information associated with the different medical user services and their presentation (application control), and also communication and service quality management data (signalling and QoS control). The developed signalling allows exchanging the characteristics associated to the different information flows between the system elements and is based on standard protocols that favour interoperability. Signalling tasks, performed by the SIP protocol, begin with the establishment of a SIP dialog with the MCU in which, by means of Session Description Protocol (SDP) messages, the different medical services are described.

The medical user services in the m-Health system are associated with information shared in a multi-collaborative environment. Specifically, the system has services to share audio, ambient video, medical data (ECG, blood pressure, heart rate and oxygen saturation), high-resolution medical still images, chat, electronic whiteboard and a web service to access clinical information databases. Each kind of information is associated with a medical user service and uses a transport protocol and a codec according to its characteristics (Table 1). Hence, real-time services (audio, video and medical data) use the Real-Time Transport Protocol (RTP) [18], whereas the rest of the services use the Transmission Control Protocol (TCP). In addition, audio and video services use the codecs recommended by the 3GPP [19], and the medical data service uses the Wavelet Transform (WT) [20].

Table 1: Characteristics of medical user services

Medical user service	Timing needs	Bandwidth needs	Transport protocol	Codec	Codec operation modes	Maximum average IPv4 bandwidth (Kbps)
Audio	RT[a]	Medium	RTP	AMR[b]	4.75-5.15-5.9-6.7-7.4-7.95-10.2-12.2 (Kbps)	28.80
Medical data	RT	Low	RTP	WT	5-10 (Kbps)	10.30
Video	RT	High	RTP	H.263	5-10 (Fps[c])	18.24
Chat	NRT[d]	Low	TCP	-	-	ABR[e]
Electronic whiteboard	NRT	Low	TCP	-	-	ABR
Still image	NRT	Medium	TCP	-	-	ABR

a. Real-Time
b. Adaptive Multi-Rate
c. Frames per second
d. Non Real-Time
e. Available Bit Rate

QoS Control Subsystem

The QoS in this system is mainly determined by the characteristics of the UMTS link. Mobile links are very fluctuant, therefore a QoS control process is required in order to obtain a good network performance. This process uses IP packet transfer performance metrics recommended by the International Telecommunication Union (ITU) in its Recommendation Y.1540 [21]. The QoS metrics selected are packet loss rate, delay variation (jitter) and octet-based IP packet throughput (bandwidth). In addition, the wireless telemedicine system has application jitter buffers to mitigate channel effects.

The QoS control process is especially important in end points because it is there where the QoS-related decisions are taken. When an end point detects that a particular communication does not operate properly, it needs to modify the characteristics of its multimedia session in order to improve QoS and thus, it renegotiates the corresponding session by sending SIP/SDP messages. Hence, the system end points and the MCU can modify certain upper-level protocol parameters (codecs used, transmission rates, compression ratios, etc.) in order to adapt the information transmitted to network performance. This process is possible thanks to a transport library that provides a uniform interface to send the information generated by medical user services and different QoS metrics measurement tools developed for several types of links. This transport library also offers different queuing policies.

Due to the variable and scarce wireless channel resources shared between all medical user services, it is necessary to prioritize them to provide an adequate treatment to real-time and non real-time ones. Real-time services are very sensitive to channel conditions (mainly bandwidth, delay, jitter and packet loss rate), whereas non real-time ones can adapt well to varying environments thanks to the built-in flow control and reliability of TCP. For that reason, the most priority services are the medical data, audio and video services, which will take up most of the channel resources. Non real-time services will be treated best-effort, adapting to the spare network resources using TCP built-in mechanisms (Table 1). According to the discussion, real-time services are the most priority, but, among them, a clinically acceptable ECG signal is more important than a clear audio conversation that, in turn, is more important than the ambient video signal. Thus, the medical data service has high priority, the audio service medium priority and the ambient video service low priority. The characteristics of these services are monitored at transmission and reception and are taken into account to increase or decrease codec rates.

Two of the main causes of poor QoS are packet losses and packet drops. Packet losses are produced inside the network, whereas packet drops occur in application queues. Regarding packet losses, they can be caused by congestion conditions or channel errors. As the commercial UMTS 3G wireless cellular service used in this system operates in the UMTS Acknowledged Data Transfer mode at the radio link layer [22], packet losses are considered to be produced only by network congestion. Thus, part of the QoS control process is based on congestion control. Congestion control signalling can be implicit, activated by packet drops in transmission queues, or explicit, initiated in reception. Implicit signalling allows controlling the congestion in the UMTS link, whereas explicit signalling is used when congestion is detected in the rest of the links in the communication path. The congestion control algorithm selected is that presented in [11], but applied to the three real-time medical user services of this system. Using codec rate adaptation according to service priorities, the codec rate of the low priority service is decreased first, and then the codec rate of the medium priority service is varied. Finally, the high priority service is limited.

Regarding jitter, it can be caused by the variable nature of wireless links or by the joint transmission of all services, therefore each real-time service has an application jitter buffer associated with it that tries to mitigate its effects (Fig. 3). These buffers have been properly dimensioned to minimize jitter, delay and packet drops. They are First In-First Out (FIFO) queues that are filled with network packets as they arrive. The application periodically empties one packet from the buffer at the required rate, but with the following considerations:

- If the buffer becomes empty, it does not serve packets to the application until a predefined buffer occupancy threshold is reached.
- From that point on, the buffer serves packets until it becomes empty. If, on the other hand, the buffer fills completely and a new packet arrives, the first packet stored in the buffer (the oldest one) is dropped to make room for the new one.

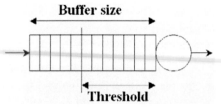

Figure 3: Application jitter buffer model

IPv4 Performance Results

In order to measure the wireless telemedicine system performance and to improve the quality of the services offered by dimensioning jitter buffers, several tests have been carried out using the system over 64/128 Kbps (Uplink/Downlink bandwidth at IP level) IPv4-only UMTS accesses in urban areas (good coverage level and low speed, as well as static vehicles).

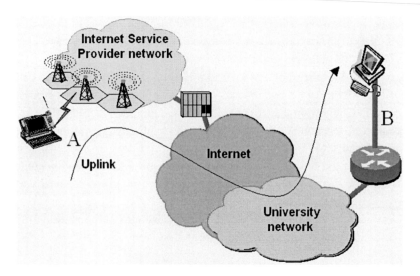

Figure 4: Measurement scenario

The measurement scenario used is shown in Fig. 4. As the uplink is more restrictive, the results presented here correspond to that connection sense. Packets have been captured in points A and B. Measurements in point A have been used to obtain the characteristics of the traffic injected into the uplink (IP-level bandwidth and jitter), whereas measurements in point B allow to obtain network behaviour (packet loss rate and jitter). Several tests have been carried out during several weeks, staggered along the whole day, all days of the week. The duration of these tests has been selected according to the average service time of an ambulance, which has been considered to be 10 minutes in a medium-sized Spanish city like Zaragoza.

Average Bandwidth Results

Table 2 presents the results about the average IP-level bandwidth used by real-time medical user services in point A. 48 tests have been carried out every 30 minutes during one day, using isolated user services and varying codec operation modes. As it can be observed, the audio service adapts well to the theoretical expected average IP bandwidth (see Table 3 and Eq. 1). In addition, considering more audio samples per network packet reduces bandwidth use, since transmission efficiency (information carried by each packet to total packet size ratio) is increased. However, there is a limit in the number of audio samples per packet that can be used because more audio samples per packet yield more audio delay (audio samples are generated every 20 ms, see Eq. 2). Moreover, if an audio packet is lost, all the audio samples carried by it are lost. Therefore, a reduced number of audio samples per packet is more suitable to error-prone environments. That is the reason why the maximum number of samples per packet has been limited to 3 in the final system, although Table 2 also shows the results for 4 and 5 samples per packet. Regarding the video service, it is worth noting that the bandwidth shown in Table 2 can vary substantially with the movement of the video scene captured. Finally, the medical data service adapts well to the codec rate specified because medical data frame sizes are long enough to obtain good transmission efficiencies.

Table 2: Average IP-level bandwidth used by real-time user services

Medical user service	Operation mode		Average IPv4 bandwidth (Kbps)
	Samples/Packet	Codec rate (Kbps)	
Audio	1	4.75	21.20
	1	12.2	28.80
	2	4.75	13.21
	2	12.2	20.81
	3	4.75	10.50
	3	**12.2**	**18.10**
	4	4.75	9.22
	4	12.2	16.82
	5	4.75	8.41
	5	12.2	16.03
Medical data	Bit rate (Kbps)		
	5		5.30
	10		**10.30**
Video	Frames per second		
	5		8.05
	10		**18.24**

Table 3: Audio sample size

Audio codec rate (Kbps)	Sample size (Bytes)
4.75	13
12.2	32

$$IPv4\ Bandwidth\ (Kbps) = \frac{8 \times (20 + 8 + 12 + Samples\ per\ Packet \times Sample\ Size)}{Samples\ per\ Packet \times 20} \quad (1)$$

$$Fixed\ delay\ (ms) = Samples\ per\ Packet \times 20 \quad (2)$$

As it can be checked in Table 2, the total bandwidth consumed by all real-time medical user services fits in a 64 Kbps UMTS channel, even when the most bandwidth-consuming codec rates and the lowest transmission efficiencies are used. Thus, according to the previous discussions, the initial codec operation modes selected in this wireless telemedicine system have been those highlighted in Table 2, achieving a reasonable trade-off between bandwidth, transmission efficiency, delay and loss rate. During normal operation, codec modes can vary in response to congestion conditions with the aid of the congestion control algorithm mentioned in subsection "QoS control subsystem". The average IP-level bandwidth obtained in point B is very similar to that obtained in A. In addition, no packet losses have been observed in any point. Therefore, the network does not modify traffic characteristics regarding bandwidth and packet loss.

Jitter Results

48 tests have been carried out every hour during 2 days, with all the real-time medical user services operating at the same time and at the codec rates highlighted in Table 2 (the highest possible codec rates). These tests are useful for observing the influences between traffics generated by each real-time service.

As all possible jitter effects can be observed in point B, Fig. 5 presents a zoom over 9 seconds of audio interpacket time taken in a test in point B. Audio packets are generated every 60 ms, so this is the theoretical time that should appear in Fig. 5. Medical data packets are generated every second (approximately every 17 audio packets), therefore their effects over the audio service appear uniformly spaced. It can be observed that they cause more than 200 ms of jitter due to the time it takes the 64 Kbps UMTS uplink channel to transmit big-sized medical data packets (about 1300 bytes at IP level). Regarding video, packets are smaller and not uniformly spaced because they depend on image movement. Thus, the effects of video over audio packets are smaller. In addition, other jitter effects are caused by the network. Finally, all the effects can overlap at the same time.

Regarding the effects of audio and video over the medical data service, none of them have a significant influence (Fig. 6) due to the fact that medical data packets are very spaced between them (1 second, ideally), and a jitter effect of less than one second is not noticed in reception with the aid of a minimal jitter buffer. This result can be checked in Fig. 6, which presents medical data interpacket time in point B for a particular test.

The last real-time service, the ambient video, also suffers jitter effects caused by the rest of real-time services (Fig. 7). However, the most serious effects are produced by the video codec used and big reception buffers are recommended. In addition, this service has the lowest priority and video motion softness is not critical, therefore a big jitter buffer is enough to support all the possible jitter effects.

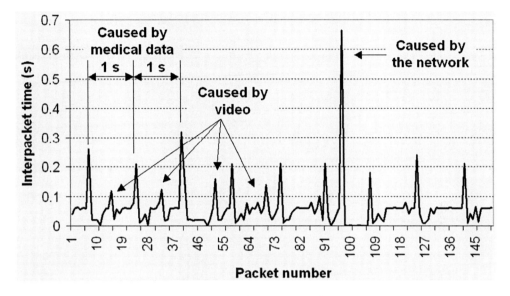

Figure 5: Audio interpacket time

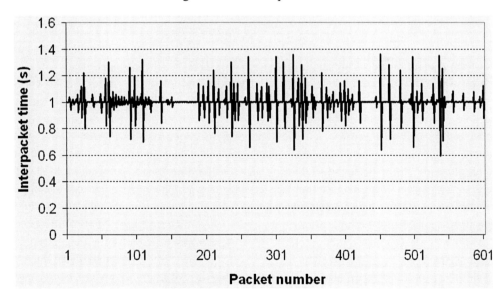

Figure 6: Medical data interpacket time

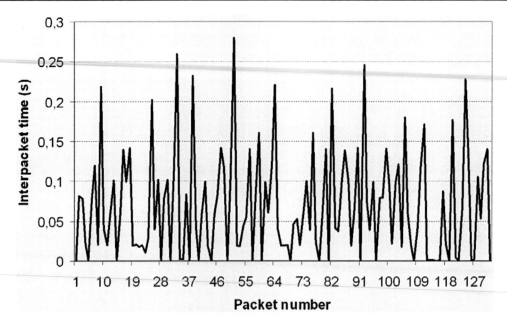

Figure 7: Video interpacket time

Jitter Buffer Dimensioning

Packet captures in point B are very useful to obtain the instantaneous application jitter buffer occupancy for all the real-time services. Using packet timestamps and theoretical packet buffer empty times, the instantaneous buffer occupancy can be calculated. Using the same tests of the previous subsection, with all the real-time medical user services operating at the same time and at the codec rates highlighted in Table 2, several buffer occupancy calculations have been carried out. In the calculations, buffer size and threshold values can be varied in order to obtain useful results to dimension the jitter buffers properly. First of all, the buffer threshold must be able to support jitter effects caused by the real-time services. Subsequently, the total buffer size must be able to support jitter effects caused by the network. High values of the threshold cause fewer situations in which the buffer becomes empty, allowing a continuous reproduction, but, on the other hand, introduce a bigger fixed delay. A value too low reduces the fixed delay, but at the expense of causing frequent interruptions in the reproduction. Regarding the buffer size, bigger buffers allow less packet drops than smaller ones, but also entail a bigger delay on enqueued packets.

Buffer Threshold

The most important jitter effects caused by other services that the audio service suffers are produced by the medical data service. Their value can vary, but rarely exceeds 200 ms. Considering that audio packets are generated every 60 ms, a threshold of 4 packets (240 ms of audio stored in the buffer) is enough in order to support them. Due to the small jitter effects present on the medical data service, the buffer threshold can be selected to be minimal. A value of 2 packets would be enough, but to ensure a more robust behaviour, and considering that fixed delay is not relevant, 3 packets has been used as the threshold for this service.

Finally, the ambient video service is not critical, so the buffer threshold selected for it has been the minimum, i.e., 1 packet.

Buffer Size

As it has been noted before, the total buffer size must be able to support jitter effects caused by the network. The first step is to consider an infinite buffer with all the 10-minute tests carried out to calculate the maximum buffer size that would have been needed in order not to drop any packet. Figs. 8, 9 and 10 show the instantaneous buffer occupancy obtained for a particular test of the audio, medical data and video services, respectively. In these cases, a buffer size of 12 packets for the audio service, 3 packets for the medical data service and 25 packets for the audio service would have been enough.

The same results have been obtained in all medical data tests (3 packets), therefore a value of 4 packets as jitter buffer size is a good choice in order to ensure a robust behaviour and no packet drops even if worse conditions appear. Regarding the video service, the maximum buffer occupancy obtained in all the tests has been 25 packets. A more conservative value of 30 packets has been selected to support even worse working conditions.

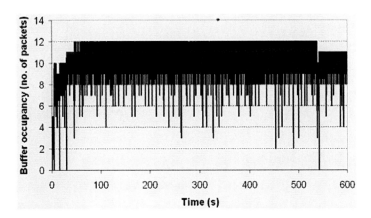

Figure 8: Audio service buffer occupancy (threshold = 4 and infinite buffer)

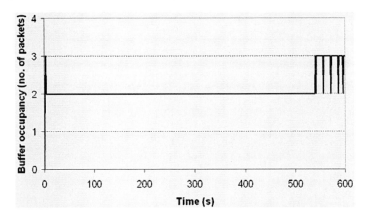

Figure 9: Medical data service buffer occupancy (threshold = 3 and infinite buffer)

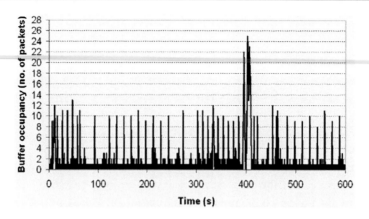

Figure 10: Video buffer service occupancy (threshold = 1 and infinite buffer)

Finally, audio tests present a small variability in the maximum buffer size. In order to obtain a suitable buffer size, finite buffer sizes have been considered, producing different packet drop ratios depending on their value. The packet drop ratio has been averaged for all the tests and the results are shown in Fig. 11 as a function of buffer size. Numeric results are also presented in Table 4. Not only packet drop ratio, but also the fixed and maximum delays that a particular buffer size causes are the relevant parameters in order to select its value. If, for example, a packet drop ratio of less than 1% is desired, 7 packets would be choice. Taking into account all the previous discussions, Table 5 presents the jitter buffer parameters finally selected for a proper operation of the m-Health system.

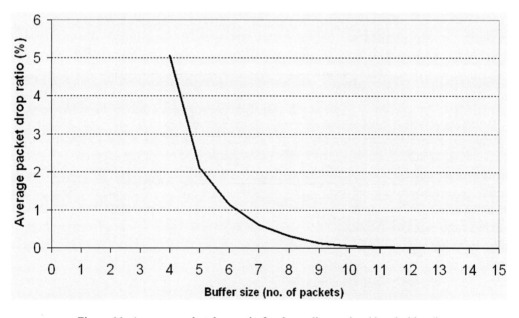

Figure 11: Average packet drop ratio for the audio service (threshold = 4)

Table 4: Average audio packet drop ratio vs. delay (buffer threshold = 4)

Buffer size (packets)	Average packet drop ratio (%)	Typical deviation	Fixed delay (ms)	Maximum delay (ms)
4	5.059	0.441	240	240
5	2.113	0.237	240	300
6	1.137	0.198	240	360
7	**0.619**	**0.153**	**240**	**420**
8	0.321	0.098	240	480
9	0.137	0.063	240	540
10	0.060	0.039	240	600
11	0.028	0.019	240	660
12	0.010	0.010	240	720
13	0.004	0.006	240	780
14	0.001	0.003	240	840
15	0.000	0.000	240	900

Table 5: Jitter buffer parameters

Medical user service	Buffer threshold (packets)	Buffer size (packets)	Fixed delay (ms)	Maximum delay (ms)
Audio	4	7	240	420
Medical data	3	4	3000	4000
Video	1	30	-	-

Conclusions

This chapter has presented the IPv4 performance of a wireless telemedicine system targeted specifically for critical and emergency medical scenarios. It offers simultaneous transmission of real-time clinical data (including ECG signals, blood pressure and blood oxygen saturation), videoconference, high-resolution still image transmission and other facilities such as multi-collaborative whiteboard, chat and web access to remote databases. Due to the nature of the services transmitted and since it is IP-based, home telecare and chronic patient telemonitoring are other application areas in which this wireless telemedicine system can be used.

The system architecture is based on 3G wireless networks and advanced signalling protocols (SIP/SDP) intended for setting up multimedia communication sessions between one or multiple clients, that also allow the integration of real-time multimedia services over multiple access channels supporting IPv4 and/or IPv6 protocols (depending on current commercial UMTS releases).

Real-time multimedia data transmission has been optimized specifically to operate over 3G wireless networks using the most appropriate codecs. IPv4 performance results show that the system performs reliably over IPv4-only UMTS accesses (64 Kbps in the uplink). The total bandwidth used fits in a 64 Kbps UMTS channel even when the most bandwidth-consuming codec rates and the lowest transmission efficiencies are used. Regarding jitter, IPv4 measurements allow dimensioning jitter buffers that improve playback quality of real-time services. The parameters selected for each jitter buffer and the design trade-offs and occupancy calculations are also presented.

The migration process towards next generation Internet and Fourth-Generation (4G) networks will require the use of the IPv6 protocol. As it has been stated before, this wireless telemedicine system is ready for IPv6, but it has been dimensioned for IPv4-only accesses so far. Therefore, the study of all system parameters and how could the IPv6 protocol affect system performance in any possible IPv4/IPv6 transition scenario will be the next step. The final objective of this work is to build a useful m-Health system ready for the current and envisioned technologies in the years to come. Finally, and taking into account current technologies, the integration of the wireless telemedicine system into a real IMS environment will also be considered in the future.

References

[1]   R.S.H. Istepanian, and J.C. Lacal, "Emerging Mobile Communication Technologies for Health: Some Imperative notes on m-Health", *Proc. of the 25th Silver Anniversary International Conference of the IEEE Engineering in Medicine and Biology Society*, Vol. 2, pp. 1414-1416, 2003.

[2]   R.S.H. Istepanian, E. Kyriacou, S. Pavlopoulos, and D. Koutsouris, "Wavelet Compression Methodologies for Efficient Medical Data Transmission in Wireless Telemedicine Systems", *Journal of Telemedicine and Telecare*, Vol. 7, No. 1, pp. 14-16, 2001.

[3]   R.S.H. Istepanian, B. Woodward, and C.I. Richards, "Advances in telemedicine using mobile communications", *Proc. IEEE Engineering Medicine and Biology Society*, Vol. 4, pp. 3556–3558, 2001.

[4]   R.S.H. Istepanian, S. Laxminarayan, and C.S. Pattichis (Eds), "*M-Health: Emerging Mobile Health Systems*", New York, Springer, 2006.

[5]   J. Laiho, A. Wacker, and T. Novosad, *"Radio Network Planning and Optimization for UMTS"*, New York, Wiley, 2000.

[6]   S. G. Miaou and C. Huang, "A next-generation mobile telemedicine testbed based on 3G cellular standard", in *Proc. IEEE Computers Cardiology*, pp. 683–686, Sep. 2001.

[7]   "IP Multimedia Subsystem (IMS); Stage 2", *3GPP TS 23.228 V7.2.0*, Release 7, 2005.

[8]   J. Rosenberg et al, "SIP: Session Initiation Protocol", *IETF RFC* **3261**, 2002.

[9]   J. Ruiz Mas, E. A. Viruete Navarro, C. Hernández Ramos, A. Alesanco Iglesias, J. Fernández Navajas, A. Valdovinos Bardaji, R.S.H. Istepanian, J. García Moros. "Design of an Enhanced 3G-Based Mobile Healthcare System", chapter 35 in the *"Handbook of Research on Mobile Multimedia"*, I. K. Ibrahim (Ed.), Idea Group Publishing, USA, 2006.

[10] E. A. Viruete Navarro, J. Ruiz Mas, J. Fernández Navajas and Cristina Peña Alcega. "Performance of a 3G-Based Mobile Telemedicine System", *Proceedings of the IEEE Consumer Communications and Networking Conference (CCNC'06)*, Vol. 2, pp. 1023-1027, Jan. 2006.

[11] Y. Chu, and A. Ganz, "A Mobile Teletrauma System Using 3G Networks", *IEEE Trans. Information Technology in Biomedicine*, Vol. 8, No. 4, pp. 456-462, 2004.

[12] J. Cullen, W. Gaasch, D. Gagliano, J. Goins, and R. Gunawardane, "Wireless mobile telemedicine: En-route transmission with dynamic quality-of-service management", *National Library of Medicine Symposium on Telemedicine and Telecommunications: Options for the New Century*, 2001.

[13] S. Pavlopoulos, E. Kyriacou, A. Berler, S. Dembeyiotis, and D. Koutsouris, "A novel emergency telemedicine system based on wireless communication technology – AMBULANCE", *IEEE Trans. Inform. Technol. Biomed*, Vol. 2, pp. 261-267, 1998.

[14] E. Kyriacou et al, "Multi-purpose HealthCare telemedicine systems with mobile communication link support", *BioMedical Engineering OnLine*, Vol. 2, No. 7, 2003.

[15] J. Wiljakka, and J. Soinien, "Managing IPv4-to-IPv6 Transition Process in Cellular Networks and Introducing New Peer-to-Peer Services", *Proc. of IEEE Workshop on IP Operations and Management*, pp. 31-37, 2003.

[16] "Interworking aspects and migration scenarios for IPv4-based IP Multimedia Subsystem (IMS) implementations", *3GPP TR 23.981 v6.4.0*, Release 6, 2005.

[17] J. Rosenberg, "A Framework for Conferencing with the Session Initiation Protocol", *IETF Internet draft*, 2004. Work in progress.

[18] H. Schulzrinne, S. Casner, R. Frederick, V. Jacobson, "RTP: A Transport Protocol for Real-Time Applications", *IETF RFC* **3550**, Jul. 2003.

[19] A. Alesanco, S. Olmos, R.S.H. Istepanian, and J. García, "A Novel Real-Time Multilead ECG Compression and De-Noising Method Based on the Wavelet Transform", *Proc. IEEE Computers in Cardiology*, pp. 593-596, 2003.

[20] "Packet switched conversational multimedia applications; Default codecs", *3GPP TS 26.235 V7.0.0*, Release 7, 2005.

[21] "IP Packet Transfer and Availability Performance Parameters", *ITU-T Rec. Y.1540*, Dec. 2002.

[22] "Radio interface protocol architecture", *3GPP TS 25.301 v6.4.0*, Release 6, 2005.

INDEX

C

D

E

India, 13, 14
indication, 259
industry, xi, 147, 153, 228
inefficiency, 118, 131
infancy, 158
infinite, 4, 207, 220, 291, 292
influence, 249, 271, 288
information exchange, vii, 63, 64, 66, 68, 104
information retrieval, vii, 159, 160, 181, 183, 185, 199
information technology, 64, 67
infrastructure, x, xi, 43, 61, 65, 66, 67, 68, 69, 71, 73, 76, 77, 80, 105, 108, 114, 170, 260, 275
innovation, ix
input, ix, 9, 10, 96, 162, 178, 184, 192, 223, 265, 266
insertion, 173, 176
insight, xi, 84, 137
instability, 239
institutions, 52
instruction, 71, 94, 95, 271
insurance, 139
integration, 64, 68, 73, 85, 103, 137, 138, 153, 154, 155, 279, 293, 294
integrity, 106
intensity, 162, 163, 181, 182, 248
interaction, 64, 68, 77, 162, 237, 260, 270
interactions, 19, 269
interest, 52, 63, 84, 89, 148, 165, 166, 197, 198, 212, 219, 220, 224, 225, 254, 260, 267
interface, viii, 27, 70, 79, 94, 105, 139, 142, 143, 148, 170, 191, 192, 199, 222, 229, 231, 235, 236, 237, 238, 239, 242, 243, 260, 272, 274, 280, 284, 295
interference, 76, 78, 79, 87, 88, 138, 139, 147, 148, 149, 153, 154, 156, 235, 237, 238, 239, 240, 241, 242, 244, 245, 246, 247, 248, 249, 250, 251, 252, 273
international standards, 148
internet, viii, 7
interoperability, 77, 139, 143, 145, 153, 282, 283
interval, 7, 252
IP address, 43, 44, 45, 46, 47, 49, 65
IP networks, 43, 117, 236, 245
Ireland, 137, 217, 218
isolation, 13
Italy, 98, 135

J

Japan, 42, 43, 60, 135, 149

K

knowledge, 23, 77, 78, 80, 152, 160, 162, 166, 198, 206, 212, 253, 271, 274, 275
Korea, 134

L

land, 260
language, 39, 78, 139, 266
laptop, 21, 63, 260
large-scale disasters, 71
latency, 48, 50, 53, 61, 120, 128, 150, 244, 264, 269, 270
lead, 110, 154, 166, 240, 242, 265, 272
leakage, 79, 262, 264
learning, 79, 167, 168, 184
leisure, ix
lifetime, ix, 122, 245
likelihood, 163
limitation, 75, 85, 93, 198, 211
links, 66, 68, 71, 93, 124, 125, 129, 133, 139, 141, 142, 145, 189, 190, 191, 239, 240, 254, 274, 284, 285
listening, 52, 204, 205
localization, 267
location, viii, ix, 20, 22, 23, 30, 34, 44, 45, 46, 50, 54, 55, 56, 58, 60, 64, 66, 70, 76, 81, 199, 200, 201, 211, 212, 227, 231, 246
location information, 46
logistics, 150
lying, 249

M

magnetic resonance, 163
magnetic resonance imaging, 163
mammal, 189
management, xi, 1, 2, 3, 5, 6, 13, 44, 49, 50, 60, 65, 66, 68, 73, 78, 108, 117, 118, 141, 142, 149, 150, 157, 158, 239, 240, 243, 247, 259, 269, 271, 273, 274, 275, 276, 280, 283, 295
manipulation, 157, 158, 159, 161, 181, 187, 193
mapping, 96, 97, 98, 100, 185, 188, 189, 227, 242, 255, 273
market, viii, ix, 2, 20, 64, 138, 139, 153, 155, 218, 260
market segment, 153
market share, 138
markets, ix, 138
Markov chain, 254
mass, ix

N

O

S

T

Y